Gruenwald
Plastics

SPE Books

Bernhardt, Computer Aided Engineering for Injection Molding
Brostow/Corneliussen, Failure of Plastics
Charrier, Polymeric Materials and Processing—Plastics,
   Elastomers and Composites
Ehrig, Plastics Recycling
Gordon, Total Quality Process Control for Injection Molding
Gruenwald, Plastics
Macosko, Fundamentals of Reaction Injection Molding
Manzione, Applications of Computer Aided Engineering in
   Injection Molding
Menges/Mohren, How to Make Injection Molds
Michaeli, Extrusion Dies for Plastics and Rubber
Rauwendaal, Polymer Extrusion
Saechtling, International Plastics Handbook for the Technologist,
   Engineer and User
Stoeckhert, Mold-Making Handbook for the Plastics Engineer
Throne, Thermoforming
Tucker, Fundamentals of Computer Modeling for Polymer
   Processing
Ulrich, Introduction to Industrial Polymers
Wright, Molded Thermosets: A Handbook for Plastics Engineers,
   Molders and Designers

G. Gruenwald

Plastics

How Structure Determines Properties

With 123 Illustrations

Hanser Publishers, Munich Vienna New York Barcelona

The Author:
Dr. Geza Gruenwald, 36 West 34th Street, Erie, Pennsylvania 16508, USA

The use of general descriptive names, trademarks, etc., in this publication, even if the former are not especially identified, is not to be taken as a sign that such names, as understood by the Trade Marks and Merchandise Marks Act, may accordingly be used freely by anyone.

While the advice and information in this book are believed to be true and accurate at the date of going to press, neither the authors nor the editors nor the publisher can accept any legal responsibility for any errors or omissions that may be made. The publisher makes no warranty, express or implied, with respect to the material contained herein.

Library of Congress Cataloging-in-Publication Data

Gruenwald, Geza.
    Plastics : how structure determines properties / Geza Gruenwald.
       p.   cm. — (SPE Books)
    Includes bibliographical references and index.
    ISBN 3-446-16520-7 (Hanser).
    1. Plastics. I. Title. II. Series.
TA455.P5G74   1992
620.1′923—dc20   92-53928

Die Deutsche Bibliothek—CIP-Einheitsaufnahme
Gruenwald, Geza:
Plastics : how structure determines properties / Geza
Gruenwald. – Munich ; Vienna ; New York ; Barcelona :
Hanser ; 1993
    ISBN 3-446-16520-7 (Hanser)

ISBN 3-446-16520-7 Carl Hanser Verlag, Munich Vienna New York Barcelona

© 1993 Carl Hanser Verlag, Munich Vienna New York Barcelona
Typesetting by The Composing Room of Michigan, Inc.
Printed and bound in Germany by Passavia Druckerei GmbH, Passau

Foreword

The Society of Plastics Engineers is pleased to endorse and sponsor this volume entitled *Plastics: How Structure Determines Properties*. SPE's Technical Volumes Committee considers this a remarkable manuscript as it presents polymer science and plastics technology from the standpoint of atomic structure. The relationship and parallelism of observed behavior to atomic microstructure provides the reader with effective structural models.

The volume is far-ranging, encompassing polymerization, physical and engineering properties, processing, composites, and structure. Consequently, it provides a long needed adjunct to polymer science texts, which are more comprehensive in some areas, but lack the breadth of subject coverage found in this book.

SPE, through its Technical Volumes Committee, has long sponsored books on various aspects of plastics and polymers. Its involvement has ranged from identification of needed volumes to recruitment of authors. An ever-present ingredient, however, is review of the final manuscript to insure accuracy of the technical content.

Technical competence pervades all SPE activities, not only in the publication of books, but also in other areas such as sponsorship of technical conferences and educational programs. In addition, the Society publishes periodicals, including *Plastics Engineering, Polymer Engineering and Science, Polymer Processing and Rheology, Journal of Vinyl Technology* and *Polymer Composites,* as well as conference proceedings and other selected publications, all of which are subject to the same rigorous technical review procedure.

The resource of some 37,000 practicing plastics engineers has made SPE the largest organization of its type worldwide. Further information is available from the Society at 14 Fairfield Drive, Brookfield, Connecticut 06804, U.S.A.

Robert D. Forger
Executive Director
Society of Plastics Engineers

Technical Volumes Committee

Raymond J. Ehrig, Chairperson
Aristech Chemical Corporation

Preface

Plastics—synthetic polymeric materials, recognized as encompassing ever-broadening properties and rapidly expanding in sheer volume of applications—can be truly understood only when compared with alternate, interchangeable materials. The similarities or dissimilarities in structure, properties, and processing conditions must also be taken into account.

As testimony to the vital importance of the selection of the right material for the fabrication of any specific item is the great number of books already written on material sciences, materials properties, and materials applications. Many of these excellent books, referred to in this book's bibliography, are, however, largely limited to a specific kind of material or written for science-oriented readers rather than materials practitioners.

During the last decades, many important perceptions have been gained in polymer science. They are well publicized in the scientific literature, but only a few of them are intelligible to readers not proficient in the disciplines of chemistry, mathematics, or engineering. The author has, therefore, attempted to make this book interesting and understandable to all individuals involved with plastics, whatever their background. Emphasis is placed on applying these findings to comprehensible solutions for common predicaments. Therefore this book contains numerous explanations and remedies for overcoming many disparities in plastics properties not revealed in the plastics literature.

Although this book is not written as a textbook, but as a compendium, its goal is to help the student, the educator, or the user of plastics materials obtain comprehensive information on mainly structural plastics materials.

It would be shortsighted, based only on the exponential growth of plastics applications, to assume that many of the old, established materials will fall into disuse. The battle between popular old established materials and innovative ones is ongoing. Therefore, one of the purposes of this book is to stress the relationship between all materials at the atomic, the microscopic, and the macroscopic levels. The reader will then be introduced to the way these structures dictate their overall properties. This, in turn, will assist the reader in choosing the most advantageous shapes, fabrications, and applications for each.

In the belief that the macroscopic properties of any material reflect the nature and position of the atoms and the strength of bonds at the atomic level, great importance is placed on conveying a qualitative understanding of the chemical nature of atoms and bonds in materials. Mathematical relationships, much elaborated on in science-oriented technical books, are left out. The reader will find explanations of all necessary chemical information, from the atoms to the macromolecules, including additives, insofar as they relate to plastics materials. The chemistry involving polymer synthesis has purposely been excluded since it is not related to the problems the plastics processor and user face.

Chemical formulas are shown consistently so they can be viewed as architectonic structures, with each stone having a particular shape and position, rather than just flat point-line connections as seen in most chemistry texts. Atomic arrangements of high strength or crystalline polymers are shown as they relate to graphite and diamond, materials of highest strength.

Compounded synthetic polymeric materials—which we call plastics—are always treated preferentially, but their relationship to natural polymeric materials—both organic and inorganic—and to metals and ceramics are pointed out. Great emphasis is placed on the changes in properties that occur on melting, solidification, crystallization, and the various aging processes, as well as responses to imposed forces that result in orientation of the polymer molecule.

As in other disciplines, scientific opinions are not always in agreement. Any author of an overview text must make personal selections. In this book, stipulations that best help the comprehension of plastics properties and that illuminate particular property implications have received favorable treatment.

The chemistry of polymerization reactions—including minute deviations—has far-reaching impact on processability and properties of materials superficially considered to be identical. However, because of the secrecy surrounding specific formulations, no accurate or helpful information can be given the plastics user. On the other hand, the literature is inundated with details on analytical chemical methods, polymer characterization procedures, and testing apparatus that are used primarily to extricate this missing information or to prove to the manufacturer of the resin that something is wrong with a specific batch of material. Both of these areas have been purposely neglected in this book.

Since the number of proprietary, brand names has multiplied over the past decade and companies are introducing a single name for a multitude of chemically different materials applied in specific areas, only customary generic or chemical names are used throughout the text to allow for unambiguous identification of any material. Also, abbreviations and acronyms are not used within the book. In the appendix, they have been gathered and explained to assist readers just scanning for information. Thoroughly indexed entries differentiate between in-depth coverage of the subject and just a cursory mentioning.

Because of the availability of a great variety of material grades, numerical property data in plastics handbooks are becoming increasingly broader. Numerical data in this book are used only sparingly, and are intended for rough comparison only. The reader is advised to peruse the manufacturer's literature for accurate data regarding a material in question. This is especially necessary for data on polymer blends, which in most cases are not even listed in handbooks.

Reluctantly, ISO recommendations for replacing the Angstrom unit by nanometer units were adopted. However, these small values will be used only for comparison purposes. In other cases where numerical values represent data more faimiliar to the reader—such as for listings of temperature and material strength properties—the old established British units or the ISO units, given in parentheses, may be chosen, depending on the preference of the reader. They need not be regarded as conversions.

August 1992   Geza Gruenwald

Contents

1
Materials

STATE OF MATTER

Early philosophers—wondering about the matter surrounding their world—correctly categorized three distinct states of matter: *gaseous, liquid,* and *solid.*

Although this book will concentrate on the latter—solid structural plastic materials—the other two have close connections to all our plastic materials. Gases are used in foamed plastics, for powering pneumatic machinery, and through polymerization or condensation of gaseous monomers for the production of a large number of commodity plastics materials. Liquids are represented by hydraulic fluids used for powering plastics processing machinery and by oils used for lubrication. Furthermore, plasticizers and other liquid additives can account for up to 40% of some plastic molding compounds. Also a great number of monomers and coreactants are liquids.

Greek philosophers added a fourth element to their understanding of the universe: *fire.* Although not considered matter, fire, as heat or the lack of it, profoundly affects the state of all materials.

Water can be used as a basic example. As a liquid, it will convert to a gas (steam) through the addition of heat. Conversely, the removal of heat converts it to a solid (ice). This kind of phase change between liquid and solid is of utmost importance to all plastics processing conditions and will be described in several chapters.

When assessing the effect of heat, several particulars have to be distinguished. As voltage and wattage (watt-hour) have quite different meanings in electricity, so one must—in regard to thermal effects on materials—distinguish between temperature readings expressed in °F (°C) or their absolute values in °R (K) and the heat energy transferred in a system, expressed in Btu (cal or J). The first one reflects the magnitude of vibration of atomic nuclei (with no motion at the absolute zero point), which, depending on other external forces (pressure and mechanical loading), will dictate the structural behavior of materials. The latter value has implications on energy requirements for the processing of plastic materials.

TABLE 1.1.   Possible Substitutions by Plastics for Established Materials, Subdivided by Their Shapes or Forms and Their Source Areas

| Basic Shapes or Forms of Materials | Inorganic Materials | | | | | Organic Polymeric Materials | | | |
| | | | | | | Natural | | Synthetic | |
	Minerals	Metals	Ceramics	Glasses	Cements	Vegetative	Animal	Thermoset	Thermoplastic
Bulk	Stones	Castings	Bricks	Castings	Castings	Wood		Yes	Yes
Slab	Plates	Plates	Tiles		Slabs	Board		Yes	Yes
Sheet	Slate	Sheet		Window panes		Plywood	Leather	Yes	Yes
Film	Mica	Foils		Flakes		Paper Cellophane	Skin	Yes	Yes
Rod		Rod		Rod			Bones	Yes	Yes
Fiber	Asbestos	Wire whiskers	Whiskers	Fibers		Cotton Hemp	Hair Wool Silk Spider web	Yes	Yes
Adhesives and coatings	Pigments	Solder Plating		Glaze Water glass	Mortar	Starch Rubber Wax Rosin Oils	Glue Casein Shellac	Yes	Yes

TABLE 1.2. Similarities between Established Materials and Plastics in Regard to Various Processing Steps

Processing Methods	Metallic Materials	Inorganic Materials	Thermoset Plastics	Thermoplastics
Compounding, alloying	Alloying	Concrete	All thermoset resins	Blends, alloys
Casting	Castings	Ceramics, glass panes	Epoxy, silicone, urethanes	Acrylics, nylon
Sintering	Powder metallurgy	Ceramics	Some thermosets	Rotational molding
Compression forming	Forging	Concrete blocks, ceramics, glass	Many thermosets	Some thermoplastics
Composite processing	Metal composites	Reinforced concrete, wire reinforced glass panes, ceramic composites	Many thermosets	Engineering and high-temperature thermoplastics
Forging, stamping	Forgings, sheet metal parts		Laminates	Some thermoplastics (mainly reinforced)
Injection forming	Die casting	Glass parts	Reaction injection molding	All thermoplastics
Extrusion forming	Aluminum tubes	Bricks, ceramics	Phenolics	All thermoplastics
Fiber drawing	Wires	Glass and carbon fibers	Few thermosets	Some thermoplastics
Rolling, calendering	Sheet and profile			Vinyl, acrylonitrile–butadiene–styrene
Laminating	Bimetallic sheets		All thermosets	Some thermoplastics
Thermoforming		Glass parts	Some thermosets	Many thermoplastic sheets and films
Blow forming		Glass bottles		Many thermoplastics

CATEGORIZING MATERIALS BY COMPOSITION

To allocate the vast number of solid materials into certain groups, they will be categorized according to chemical source and separated into groups of naturally occurring or man-made substances.

SHAPES AND STRUCTURES OF MATERIALS

Since the mechanical properties of bulk materials are not necessarily representative of the properties of specially formed or shaped products, Table 1.1 tabulates a great variety of materials for each subgroup. It illustrates how plastic materials have been and continue to be developed both as substitutes for and as improvements over some of the other established materials.

Table 1.2 illustrates some of the processing similarities between new plastic materials and other established older materials.

REFERENCES

M.F. Ashby and D.R.H. Jones, Engineering Materials. Pergamon Press, Oxford, 1980.
R.M. Brick, A.W. Pense, and R.B. Gordon, Structure and Properties of Engineering Materials, McGraw-Hill, New York, 1977.
J.E. Gordon, The New Science of Strong Materials, 2nd ed. Princeton University Press, Princeton, NJ, 1976.
J.E. Gordon, Structures, or Why Things Don't Fall Down. Plenum Press, New York, 1978.

2
Atomic Building Blocks

Just enough information about the atomic composition of structural materials will be given in this chapter to understand the interatomic and intermolecular forces. These forces, in turn, are reflected in the mechanical properties exhibited by any one of the great variety of structural materials.

PERIODIC SYSTEM OF ELEMENTS

The smallest particles that make significant contributions to the weight or mass of all materials are the atoms. Although more than 100 distinct species of atoms are known, only one-third are of technical importance and of these only eight are represented in common plastics. All atoms fit into a well-regulated order when plotted according to increasing atomic weight and arranged in subordinated columns of atoms with similar chemical properties. They are presented in an all-encompassing tabulation called the periodic system of elements. In Figure 2.1 the first 18 of these elements are illustrated, arranged in a counterclockwise spiral to emphasize the sequential pattern existing among all elements. This presentation would also provide the needed space to accommodate the growing number of atoms crowded in the consecutive four spiral turns of the remaining atoms (not shown on drawing). This missing outer part of the spiral consists (with very few exceptions) of metals, including the very important metals titanium, iron, copper, and their alloying constituents.

FIRST EIGHTEEN LOW ATOMIC WEIGHT ELEMENTS

Hydrogen

The first element in the center, the *hydrogen* atom, has an atomic weight of one. Atomic weights represent relative weights used to compare the mass of one element with that of any other element. The hydrogen atom is composed of one proton and one electron. The proton has a mass of one and carries a positive electric charge of one and the electron, which has only an insignificant mass, carries an equally large but negative electric charge.

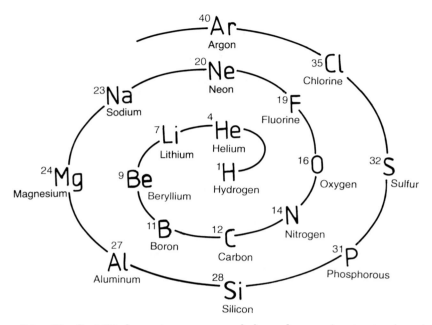

FIGURE 2.1.   The first 18 elements: names, symbols, and approximate atomic weights.

Hydrogen atoms will spontaneously react with each other forming a hydrogen molecule, which consists of two protons and two electrons. It should be assumed that the two protons (nuclei) are arranged at a fixed distance of about 0.06 nm (1 nm = 4×10^{-8} in. or 1×10^{-9}m). They are held together by the two electrons rapidly encircling them in an elliptical orbit, but not necessarily in one plane as is usually shown in graphs.

Hydrogen can combine with nearly all elements shown on the chart. The most important compounds for our consideration are those formed with carbon. With the exception of a few fluorocarbon polymers, all plastics contain hydrogen. The amount of hydrogen contained in polymers is closely related to their flammability, since the bonded hydrogen atoms can readily form combustible gases during thermal decomposition.

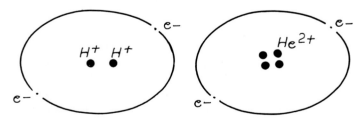

FIGURE 2.2.   Atomic models of hydrogen and helium gas.

Helium

Helium, the next element in the spiral, consists of two protons, two neutrons, and two electrons. In Figure 2.2 the arrangement of the mass and electrical charges carrying particles of the hydrogen molecule and the helium atom is compared. In helium, two of the four heavy dots represent neutrons, which carry no electrical charge; the other two are protons, which contribute not only to the mass but also to the two positive charges of the helium nucleus. Both protons and neutrons have equal mass.

A certain similarity is evident. Both elements represent practically useful gases, which are considerably lighter than air. Hydrogen can react with oxygen in the air forcefully when ignited, forming water as a combustion product. Helium, on the other side, a so-called noble gas, shows similarities to the heavier atoms neon and argon above it on the drawing, which are all chemically nonreactive.

Lithium, Beryllium, Boron

The next elements, *lithium, beryllium,* and *boron,* are all metals of low atomic weight. In addition to their inner shell electron pairs, their nuclei are orbited by one, two, and three electrons, respectively, in the surrounding, outer electron shell. Only the electrons in the outer shell partake in any chemical reaction. This shell will be fully occupied only when eight electrons are orbiting at that energy level. The element with the full compliment of electrons is *neon,* also a noble gas.

Beryllium metal is alloyed with copper for plastic molding dies; boron, in the form of borates, is an additive for plastics used to diminish their flammability. High modulus boron fibers, manufactured by a complicated process, serve as reinforcements in some advanced composites.

Carbon

The next element, *carbon,* occupies a very special position. It is situated on the center line of the periodic system of elements. Its nucleus consists of six neutrons and six protons (bearing six positive charges) and is surrounded by four valence electrons in the outer and two shielded electrons in the inner shell (Fig. 2.3).

The significance of the number four is that it is exactly between 0 and 8. This center position results in a perfect balance between negative and positive charges with all bonding electrons being truly centered. Only the carbon atom exhibits this ideally covalent character, facilitating the nearly unlimited stringing and branching of like carbon atoms. Furthermore, the spatial orientation of the valence electrons also coincides with the spatial coordination of the four neighboring atoms. In most ionic chemical compounds the number of coordinations is larger than the number of chemical valences. This concordance entails that carbon atoms are usually surrounded by four atoms placed

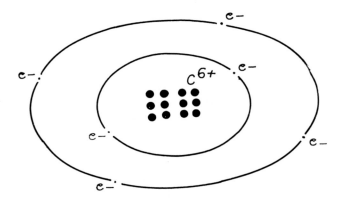

FIGURE 2.3.   Atomic model of carbon.

in space at the corners of a tetrahedron. In this fashion one carbon atom can bond to one to four other carbon atoms. These, in turn, could also bond to additional carbon atoms to form long chains (thermoplastics), endless networks (thermoset plastics), and ultimately crystalline diamond. No other element possesses this capability since only with carbon are all bonding electrons centrally located between all carbon atoms. The number of possible combinations is nearly endless, and comprises all compounds of interest to the organic chemist. Plastics, which consist mainly of polymeric carbon-based compounds, represent only a minute segment of all organic compounds.

But carbon is not restricted to organic compounds. It is widely found in inorganic compounds such as carbonates and carbides. The carbonates are widely used as fillers in plastics in the form of finely ground or precipitated calcium carbonates. Carbides decreased in importance when the synthesis of monomers was switched from the raw material acetylene (made out of calcium carbide) to petrochemicals.

Atomic Structure of Diamond, Lonsdaleite, and Graphite   The three elemental crystalline forms of carbon should be described in detail since they bear certain close relationships to the atomic structure of crystalline plastic materials in particular. These are *diamond,* the diamond-like *lonsdaleite,* and *graphite.* Their atomic structures are shown in Figure 2.4, which illustrates their structural differences.

Diamond is the hardest material known, registering a hardness of 10 on the Mohs hardness scale. It also has the highest modulus of elasticity and a very high tensile strength, but is quite brittle. It can be optically clear, colorless, and transparent, having great luster. It does not conduct electricity but is an excellent conductor of heat. All these properties are the same in all three spatial directions. Lonsdaleite is a rarely occurring variety of diamond with nearly the same properties but somewhat different structure.

Graphite represents the softest mineral, registering a hardness of 1 on the Mohs hardness scale. Its low hardness is the result of its lamellar plate struc-

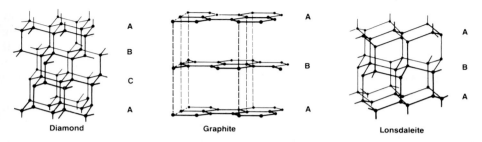

FIGURE 2.4.   Atomic structure of diamond, lonsdaleite, and graphite. Reprinted with permission from Chemical & Engineering News, *67* (20), p. 27 (1989). Copyright © American Chemical Society.

ture, which very easily shears when stresses are applied. It is optically opaque and can appear black or gray depending on the reflection of light. It is an electrical conductor.

In the diamond crystal, each carbon atom is surrounded by four neighboring carbon atoms at a distance of 0.154 nm, which are spread at 109.5° angles between any of these four directions. This tetrahedron arrangement represents a very regular, highly symmetric structure that resists deformations from external forces until those forces are high enough to split the crystal, preferably at certain planes.

In graphite, on the other hand, each carbon atom is surrounded by only three neighboring carbon atoms at a distance of 0.142 nm. Since they are all in one plane the angle is exactly 120°. The adjacent planes are spaced at a considerably greater distance, 0.339 nm. Since all carbon atoms have a valence of four (the four electrons in the outer shell), one of the three bonds in the plane must be assumed to be a double bond. The distance between two adjacent atoms gives an indication of the bond strength. The 0.339 nm spacing separating the hexagonal graphite layers is too great for a covalent bond and is responsible for the flaking of graphite and its low strength. The planar arrangement of the carbon atoms and the presence of the resonant double bonds (surplus of electrons) are responsible for the opaque dark color and the electrical conductivity. These features are also present in most electrically conductive polymers.

Resonance denotes a condition in which an additional bond is not localized to a specific bond but must be assumed to belong to several bonds within brief sequential time intervals, resulting in atomic distances that are between the standard distance and the double or triple bond distances. The very short distance between carbon atoms within the graphite plane (0.142 nm) indicates that the graphite carbon bonds are more stable than those of diamond (0.154 nm). At high temperatures, diamond can be converted to graphite; the reverse is possible only at extremely high pressures.

Although bulk graphite has very low strength, the high bond strength of the carbon bonds in the hexagonal planes bears the potential for an extremely strong and high modulus substance. The problem lies in obtaining large enough planes and extending them in a particular direction. A costly, slow process that converts carbon-containing fibers into carbon fibers exhibiting the graphite structure—in which the strong carbon bonds are aligned (albeit not straight) in the fiber direction—has made it possible to utilize this strength. The extent to which that alignment has succeeded determines the strength and modulus of a particular fiber grade.

The features of the carbon–carbon bonds in these three materials reoccur in organic compounds, especially the strong polymeric organic materials emphasized in this book. It will be repeatedly shown how many of these semi-crystalline materials can be visualized as being carved wholly or in segments out of these three carbon modifications.

The emphasis of this relationship may seem to be excessive, but it is used in this book since it makes it easier to comprehend the conformations of many structural polymers, e.g., the geometric arrangement of the atoms along a polymer chain.

Nitrogen, Oxygen, Fluorine

The next three elements, *nitrogen (1a)*, *oxygen (1b)*, and *fluorine (1c)*, have more than four electrons in their outer electron shell. Therefore when they are forming chemical bonds, they accept electrons from their reaction partners to bring the number of electrons in their own outer electron shells to eight, rather than donate electrons to other elements. Ordinarily the valences are three for nitrogen, two for oxygen, and one for fluorine. All three elements form molecules containing two atoms. Each dash stands for two electrons. The outer electrons not involved in a covalent bond are usually omitted from chemical formulas but are shown here as dashes at the sides of the atoms.

$IN{\equiv}NI$	$O{=}O$	$IF{-}FI$
N₂	O₂	F₂
Nitrogen Gas	Oxygen Gas	Fluorine Gas
(1a)	*(1b)*	*(1c)*

Nitrogen and oxygen are the main constituents of air. Fluorine is one of the very reactive gases used in the plastics industry to surface react with poly-olefin containers to reduce the vapor transmission rates of those plastics (automobile fuel tanks).

All three elements are extensively introduced into monomers to impart special properties to the resulting plastics. Their effect will be more thoroughly described later; just the main features will be mentioned here.

Ammonia (NH_3) is one of the simplest nitrogen-containing compound. This very reactive gas, which exhibits basic or alkaline properties, will impart these properties to many polymers that contain this three-valent nitrogen structure. Many of them exhibit a lower resistance toward acids, bases, and their solutions. Nitrogen- as well as oxygen-containing polymers have greater sensitivity toward moisture than the purely hydrocarbon- or halogen-containing hydrocarbon polymers.

Since the primary bond strength represents the single most important influence in regard to thermal resistance of any material, the fluorocarbon polymers have a high thermal stability. From Table 3.2 (p. 23) one recognizes that the C—F bond energy (439 kJ/mol) is higher than the C—H bond energy (414 kJ/mol). This difference of 25 kJ/mol is increased by another 70 kJ/mol in those instances in which two fluorine atoms are bonded to the same carbon atom, as is the case for polytetrafluoroethylene. With increasing fluorine content, these polymers turn into compounds with high melting points, excellent solvent and oil resistance, low gas and vapor permeability, low coefficients of friction, and superb chemical resistance. Their surfaces will also become increasingly nonwettable.

The last 9 considerably heavier elements of the 18 elements in Figure 2.1, starting and ending again with noble gases (*neon* and *argon*), carry the same number of chemically reactive electrons in their outer valence shell as the eight elements described to this point. However, in this series, the outer shell represents the third electron shell.

Sodium

Sodium, a very reactive metal, can be present in some ionomer resins. Small amounts of sodium or other alkali metal salts may remain as impurities in some polymers originating from catalysts or reactive by-products.

Magnesium, Aluminum

The next two elements *magnesium* and *aluminum,* also metals, are used to build low cost molds and molds for parts needed for low production volumes. Their compounds have found extensive use—magnesium hydroxide in glass fiber–reinforced reactive polyester compounds and aluminum trihydrate as an excellent compounding ingredient for plastics to increase arc track resistance and flame retardancy and to lower smoke generation. Metallic aluminum is the most common metal used for metallizing plastic parts.

Silicon

The *silicon* element, with its four electrons in the outer valence shell, should resemble the carbon atom in capability. The presence of the various silicone polymers will seem to confirm the ability of the silicon atom to form polymeric structures as well. This, however, is not quite the case. Silicon, as a nonmetallic element, has acquired great importance as a semiconductor material, but the silicones are more related to silica (silicon dioxide), a compound consisting of one silicon atom and two oxygen atoms. It is found in the form of quartz crystals in many stones and sands. It can also be encountered as an amorphous glass when a silica melt is quickly cooled. Crystalline silicates are of even greater importance since they naturally occur in various atomic configurations. In many instances they are applied as reinforcements or fillers in plastics compounds. A new application for silica exists in the form of a very thin (50-nm) electron beam surface deposition onto plastics films or containers to improve their barrier properties.

Tables 1.1 and 1.2 in Chapter 1 (pp. 2 and 3) demonstrate the importance of many inorganic materials for structural applications. To emphasize the relationship between the carbon-based plastic materials and the silica-based mineral and glassy materials, some of the structural similarities should be pointed out here.

Atomic Structure of Silica-Derived Polymeric Materials   Although silica occurs in quite different forms, the silica with the highest melting point, cristobalite, will be considered first. In its idealized form, it crystallizes in exactly the same crystal system and habit as diamond. The silicon atoms occupy all the positions of the carbon atoms in the diamond crystal but they are extended nearly twice the distance since an oxygen atom is inserted into the center of each bond.

Figure 2.5 shows the diamond crystal structure at the left and, at the same magnification, the idealized cristobalite crystal structure at the right. In diamond each carbon atom is surrounded by four carbon atoms. In cristobalite each silicon atom is surrounded by four oxygen atoms also in the shape of a tetrahedron, but each oxygen atom is flanked by only two silicon atoms. The chemical formula is therefore SiO_2.

The large voids in the structure of cristobalite must be reflected in property weaknesses since macroscopic properties always derive from structural features at the atomic level.

	Diamond	Cristobalite	Quartz
Specific gravity, g/cm³	3.51	2.32	2.65
Hardness, Mohs scale	10	6.5	7
Fusion temperature, °F (°C)	6920 (3827)	3115 (1713)	2658 (1459)

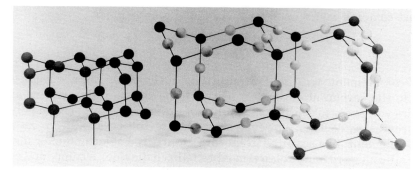

FIGURE 2.5.   Atomic structure of diamond (left-hand side) and silica (cristobalite) shown at the same scale. Black balls represent carbon or silicon, white balls oxygen. (See also the stereo picture in Appendix C.)

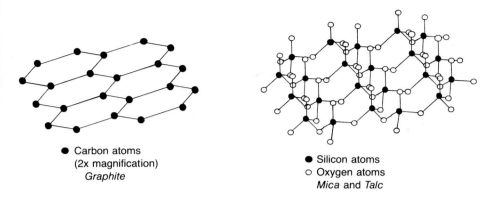

Carbon atoms
(2x magnification)
Graphite

Silicon atoms
Oxygen atoms
Mica and *Talc*

FIGURE 2.6.   Atomic structure of graphite and mica. Left: Graphite, Right: Mica, and talc.

In spite of the much higher atomic weights of silicon (28) and oxygen (16) in comparison to carbon (12), the specific gravity of the silicas is much lower. This spaciousness of the silicas reappears again with the silicone polymers. It becomes expressed not only in their relatively low specific gravity but mainly in their extremely high diffusion rates for gases and water vapor.

Just as most plastic materials can be pictured by the arrangement of their carbon atoms in the diamond or graphite atomic crystal lattices, so can the silicas and silicates be derived just by substituting each carbon atom by a silicon tetrahedron $[SiO_4]^{4-}$. Other silicas with lower melting points, which include common quartz, can be visualized by crumpling the cristobalite structure somewhat, in which case the Si—O—Si angle changes from 180° to only about 145°. These less symmetric forms still retain the identical tetrahedral structure around each silicon atom.

Applying the same criterion of graphite to a silicate, one obtains the crystal form of mica, a material also representing a layered structure. In Figure 2.6 the structures of both materials are compared. The distances between carbon atoms are shown doubled so that the carbon and silicon atoms will appear at the same relative points. Not to overcrowd the picture, only carbon, silicon and oxygen atoms are shown. Since the charges of the silicon and oxygen atoms in this arrangement do not balance out, additional positive ions must be added (such as potassium and aluminum). This structure is actually more complicated because of the substitution of some silicon atoms by aluminum atoms. However, of importance for the mechanical properties are the double silicate layers corresponding to the formula $[Si_4O_{10}]^{4-}$.

Mica, as well as talcum, is a mineral displaying a sheet-like structure. No corresponding arrangements of atoms are known in organic polymeric materials. Only the surface layers of thermoset resins could be visualized as having similar characteristics.

The same principle can be extended to analogous fibrous structures such as carbon or graphite fibers and asbestos fibers. In this instance, the structure of asbestos is well known; however, very little is known about the carbon fiber structure. The basic asbestos fiber must be assumed to consist of two strands of

• Carbon   • Silicon   ○ Oxygen

FIGURE 2.7. Atomic structure of anthracene (left), graphite (middle), and asbestos (right) fiber (plan view with only chain atoms shown at about the same scale).

cumulated, silicon atoms carrying, hexagon rings. A low-molecular-weight carbon analog would be anthracene, as shown in Figure 2.7.

To obtain a better picture of these two silicate substances, the structures of their cross sections should be shown for comparison. Again, the structure's other cations have been omitted in Figure 2.8 to illustrate the essential features.

Since asbestos fibers are composed of atoms of several different atomic weights, not just one as in the case of graphite, electron microscopic pictures of asbestos fiber cross sections will show the fiber's structure in good contrast (Fig. 2.9).

The circular or spiral arrangements in the asbestos fibrils enlighten the supramolecular structure. Although the chemical structure would indicate that each fiber element occupies a rigid rectangular space, having a thickness

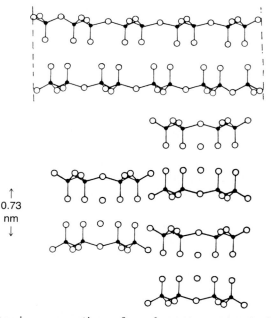

0.73 nm

FIGURE 2.8. Atomic cross sections of an elementary mica double layer (top) and of three elementary asbestos fiber molecules (bottom) (black dots silicon and circles oxygen).

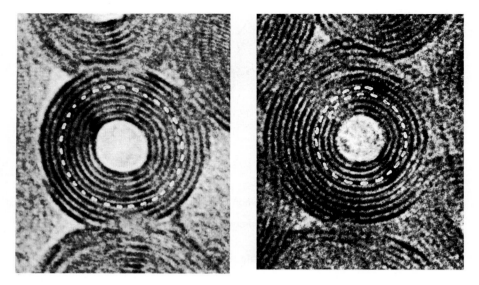

FIGURE 2.9. Transverse section of chrysotile asbestos fibrils, having a diameter of about 20 nm. From K. Yada, Acta Crystallogr. 26, 659 (1971).

of 0.73 nm as seen in Figure 2.8, consecutive rectangles may actually be curved to make up the diameter of those circular fibrils (20 to 30 nm). The convergence of these rectangles, seen as melding black dots in the photos, forming the circular lines, can also explain the structure of graphite fibers.

Since about half of all plastic materials do not exhibit a well-ordered crystalline structure, the corresponding amorphous silica analogs also need to be considered. In the idealized cristobalite the silica tetrahedrons, regardless of the direction from which they are observed, are arranged as symmetric six-membered rings. In silica glass the tetrahedrons are arranged in four- to eight-membered irregularly kinked rings, arranged in irregular planes. An even greater irregularity in the arrangement of the tetrahedrons is obtained when other cations (sodium, calcium, boron) are introduced, as is the case for ordinary glass. As a result, the substance's atomic structure is altered from that of quartz glass to that of ordinary glass by the unshackling of some of the —Si—O— rings caused by the interstitial positioning of the other metallic ions. Those three-dimensional networks are structured very similarly to the urea-formaldehyde resins. Pictures of the atomic arrangements for glass or thermoset plastics are only sparingly given in this book since they are all misleading. To visualize the kind of disposition in these systems one should imaging dropping a long chain into a cubical container and then connecting some of the chain members in irregular intervals with adjacent members by means of additional short chain links. The resulting random positioning of any chain link reflects well the actual structure of many such materials.

It is important to realize that this irregular, tightly cross-linked structure cannot be converted into an oriented structured shape by drawing or rolling;

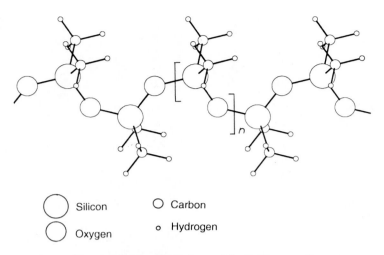

○ Silicon   ○ Carbon

○ Oxygen   ○ Hydrogen

FIGURE 2.10.   Atomic model of silicone oil.

therefore silicate glasses remain isotropic, whether drawn into fibers or cast into sheets. The same is true for thermoset plastics. The strain improved higher tensile modulus of elasticity observed on high-modulus, high-temperature phenolic fibers is the result neither of fiber orientation nor crystallization, but may just be caused by an improved alignment of the benzene rings.

The rudimentary polymeric silicones consist of silicon compounds that bear two strong Si—O bonds that form the extended chain structures found in both liquid and elastomeric silicones. An atomic model of such a chain is shown in Figure 2.10.

The highly rigid silicones contain an increasing number of trifunctional Si—O bonds that form interconnected ring structures. The remaining two bonds, or only one bond, respectively, are satisfied by methyl radical groups. All four bonds surrounding the silicon atom are not flat but, like carbon, are tetrahedral in shape. The bonds in hard silicone resins could roughly be presented as shown in Figure 2.11.

If Figure 2.11 looks confusing, the reader is getting the right impression. By following all Si—O bonds, the two possible link patterns, O—Si—O and

O   O can easily be discerned.
  \   /
   Si
   |
   O

The great variety of silicone plastics and rubbers are obtained not only by utilizing these various functionalities of the silicone atom but also by partially substituting some methyl groups with vinyl, phenyl, fluoroalkyl, or other radicals. High vinyl functionality will yield higher cross-link densities, resulting in better solvent resistance and compression set resistance. Increased fluorine content improves ozone and chemical resistance to gasoline, solvents, and oils and enhances thermal stability.

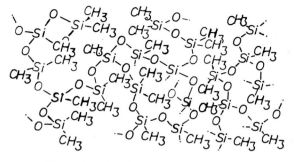

·— Out-of-plane bonds

FIGURE 2.11.   Arrangement of atoms in hard silicone resins.

Low-molecular-weight, reactive, substituted silanes are also widely used in compounded plastics since they—if properly selected—act as adhesion promoters. One-half of the molecule should chemically bond to the surface of inorganic fillers or fibrous reinforcements and the other half should become incorporated into the polymer stock.

On the other hand the nonreactive silicone fluids are used as mold releases and as antifoaming additives. Their presence often causes bond failures in plastic parts that are painted or bonded with adhesives.

Phosphorus

The next element of increased atomic weight is *phosphorus,* which is a member of the nitrogen group. It can have a valence of three, but most compounds found in plastics derive from the five-valent phosphorus atom. Phosphate esters were originally used as plasticizers for polyvinyl chloride compounds. They are not only good plasticizers but also contribute, like many other phosphorous containing formulations, to their flame-retardant properties. Phosphite compounds are finding application as thermal stabilizers for plastics.

Sulfur

Next comes *sulfur,* a member of the oxygen group. Although oxygen is always two valent, sulfur in compounds occurs at a great variety of valences. It is an important element in many natural and synthetic polymers. Three classes of plastics contain sulfur as a component in the polymer chain *(2, 3, 4):*

1. the aliphatic polysulfides *(2),* which represent the most solvent and chemical resistant elastomers;
2. the aromatic polysulfides (mainly polyphenylene sulfide *(3)*), which are also chemically resistant but in addition withstand high temperatures; and
3. the polysulfones and polyether sulfones *(4),* which represent high-temperature, transparent thermoplastic engineering resins.

The sulfur atom in the first two categories bears two or four binding electrons. The sulfur atom can directly or—via additional sulfur atoms—indirectly be bonded to carbon atoms on each side. The plastics of the third group contain the six-valent sulfur atom, which also has two oxygen atoms bonded at the sides to the chain linking sulfur atom.

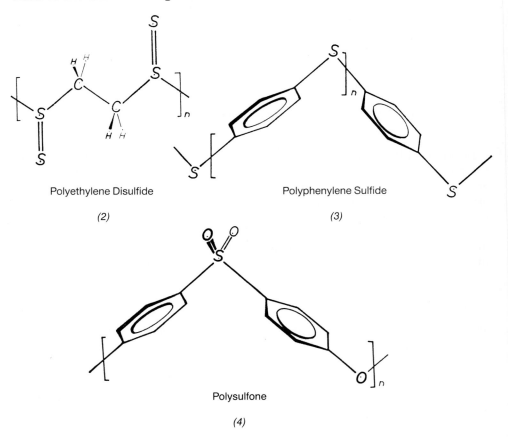

Polyethylene Disulfide

(2)

Polyphenylene Sulfide

(3)

Polysulfone

(4)

Another important use for sulfur is its application as a vulcanizing agent for rubber. There several sulfur atoms (marked with black dots in Fig. 2.12) in a row connect adjacent polymer chains by cross-links.

The property improvements contributed by the sulfone groups are further utilized in the manufacture of chlorosulfonated rubber. Although the incorporation of sulfone groups by means of a surface treatment in polyethylene tanks can promote the barrier properties for petroleum fuels, their predominance has shifted to other processes.

Chlorine

The next atom is *chlorine*, which carries seven electrons in its outer valence shell, the same as fluorine and bromine. For its application in all plastic

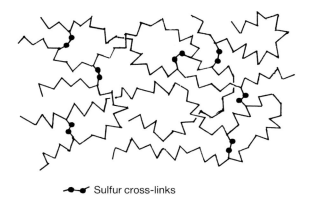

Sulfur cross-links

FIGURE 2.12.   Sketch of cross-links in vulcanized rubber.

compounds, single valency persists. Like fluorine, it can substitute for hydrogen atoms in any monomer. However, generally only one or two of these are exchanged with chlorine. These polymers have good solvent and oil resistance, excellent chemical resistance, and are usually flame retardant. However, chlorine and other halogen-containing polymers may, under improper processing conditions, prolonged exposure to high temperatures, or when involved in combustion processes, form highly corrosive acids, such as hydrochloric acid.

The subsequent halogen element in the fluorine group is *bromine*. It is positioned beyond the spiral shown in Figure 2.1, at the upper right corner. Bromine is sometimes preferred over chlorine for incorporation in flame-retardant formulations because of its greater efficiency based on its high atomic weight (about twice that of chlorine).

REFERENCES

I. Bernal, W.C. Hamilton, and J.S. Ricci. Symmetry. W.H. Freeman, San Francisco, 1972.

R.M. Brick, A.W. Pense, and R.B. Gordon, Structure and Properties of Engineering Materials, McGraw-Hill, New York, 1977.

M. Grayson, ed. Encyclopedia of Glass, Ceramics, Clay and Cement. John Wiley, New York, 1985.

H.D. Megaw, Crystal Structures, Working Approach. W.B. Saunders Co., Philadelphia, 1973.

L.J. Monkman, Asbestos—recent developments. In Applied Fibre Science, Vol. 3, p. 163. F. Happey, ed. Academic Press, New York, 1979.

3

Variety of Interatomic Bonds

Interatomic relationships are of great importance for the characterization of plastic materials since they will also determine their properties. Each type of chemical bond is subject to certain rules and, although quite different in nature, may coexist with or influence other bonds under certain conditions. The curve in Figure 3.1 basically describes one common point in relation to the dependence of these forces on the distance between atoms. The crossing point of the curve with the abscissa indicates the distance between two atoms under normal conditions. The steep curve extending to the positive y axis shows the rapidly increasing repulsive force, which is indicative of the great difficulty in compressing solid matter. The shallower curve extending to the negative y axis registers the minimum energy required for breaking the bond. Once the bond has been broken, less energy is required to separate the atoms completely.

COVALENT BONDS

Among the four classes of bonds, the *covalent bond* is of prime importance for all organic chemicals and thus also for the organic polymeric substances forming all plastic materials. Some comparative numerical values are listed in Tables 3.1 and 3.2.

   The covalent bond is characterized by two atoms in such close proximity that one electron from each atom's outer shell will orbit both nuclei jointly as an electron pair. If two atoms are joined by the formation of two electron pairs, the resulting bond is called a double bond. Since individual atoms have different numbers of bonding electrons in their outer shell, they are capable of forming a corresponding number of bonds, but the limit is four for covalent bonds. The atomic structures of diamond and lonsdaleite represent the only two possible spatial arrangements for a condition in which each carbon atom is equally bonded to four other carbon atoms.

   This simple valence-bond theory cannot correctly explain all facts relative to chemical bonds, therefore a more complicated molecular-orbital theory has been established, based on a mathematical description of molecules carrying different kinds of bonds. For the cursory considerations in this text, the simple valence-bond theory will suffice as long as one realizes that a double bond is

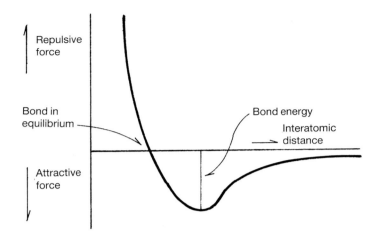

Figure 3.1.   Molecular bond energy as a function of the distance between two nuclei.

not the same as two single bonds. Likewise, because of resonance, a single bond and a double bond originating from the same atom should not necessarily be considered static as marked by the dashes in the outlying formulas. The compound's behavior will be more representative of a compound containing partial bonds as illustrated in the center. This bond should be considered oscillatory rather than stationary in one or the other position.

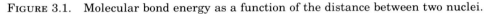

Electron bond resonance

(1)

Table 3.1.   Atom Radii (nm)

Atoms		Valence Radii			van der Waals Radii
Symbols	Name	Single Bond	Double Bond	Triple Bond	
H	Hydrogen	0.032	—	—	0.12
C	Carbon[a]	0.077	0.067	0.060	—
O	Oxygen	0.066	0.059–0.062	—	0.14
N	Nitrogen	0.070	0.063	0.055	0.15
F	Fluorine	0.064	—	—	0.135
Cl	Chlorine	0.099	—	—	0.18
S	Sulfur	0.104			
Si	Silicon	0.117			

[a]Distances between carbon atoms in diamond, polyethylene, and cyclohexane, 0.154 nm; graphite, 0.142 and 0.339 nm; and benzene 0.142 nm.

TABLE 3.2.   Covalent Bond Data

Atom Bond	Distance between Atoms (nm)	Bond Energy (kJ/mol)	Strings of Atoms	Bond Angle (°)
C—C	0.154	347	C—C—C	109.5
C⋯C Aromatic	0.139	410	C⋯C⋯C	120
C=C	0.135	614	C=C—C	125
C≡C	0.120	811	C≡C—C	180
C—H	0.109	414	H—C—H	109.5
O—H	0.096	460	H—O—H	104.5
O⋯H Hydrogen bond	~0.25	12 to 25		
N—H	0.101	389	H—N—H	107
N⋯H Hydrogen bond	~0.30	12 to 20		
C—O	0.143	351	C—O—C	108
			C—C—O	110
C=O	0.123	715	C—O—C (‖ O)	113
			C—C—O (‖ O)	116
C—N	0.147	293	C—N—C	108
			C—N—C (‖ O)	117
			C—C—N (‖ O)	113
C=N	0.127	614		
C≡N	0.116	890		
C—F	0.149	439		
C—Cl	0.177	326		
Si—O	0.164	368	Si—O—Si	142
			O—Si—O	110
Si—C	0.187	288		

The simplest example of a covalent chemical bond is the bonding of two hydrogen atoms forming one hydrogen molecule. In analogous fashion, the same kind of bond is encountered in molecules containing more than two atoms and molecules containing different atoms. Using four basic compounds, rudimentarily incorporated and repeated in many plastic materials, the characteristics of these covalent bonds may be clarified:

	Molecule			
	Methane	Ammonia	Water	Hydrogen Chloride
State of matter[1]	gas	gas	vapor	gas
Chemical formula	CH_4	NH_3	H_2O	HCl
Molecular weight	16.0	17.0	18.0	36.5

The molecular weight of these molecules can easily be obtained by adding the atomic weights of all the elements present in the molecule, e.g., for one carbon atom 12 and for four hydrogen atoms 4, resulting in 16 for methane.

Each of these compounds contains hydrogen atoms with a valence of one, e.g., the ability to form only one covalent chemical bond. The other elements, which have more than one electron in their outer shell, are able to form more than one bond. To illustrate the disposition of bonds (shown with dashes), the chemical formulas (2) are as follows:

(2)

These drawings represent a very convenient way of expressing the local positions of all elements in a molecule and are therefore almost always used when every bond in a molecule must be described. However, since many atoms contain more electrons in their outer shell than those participating in covalent bonds, a complete picture can be obtained only by including all outer shell electrons (each electron is expressed as a dot). The outer electron structure of these elements would then be shown as (3):

$$\dot{\underset{.}{\overset{.}{C}}}\cdot \qquad \dot{\underset{..}{N}}\cdot \qquad \overset{..}{\underset{..}{O}} \qquad \overset{..}{\underset{..}{C}l}:$$

(3)

The corresponding compounds can be rewritten on a flat piece of paper to give (4):

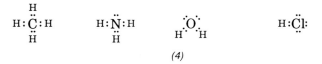

(4)

---

[1] The reader must clearly distinguish the materials listed as gases or vapors from the commonly available liquids sometimes bearing the same name. These liquids are either associations of molecules formed by secondary bonds (hydrogen bonds), as in the case of water, or are solutions of these gases in water, as in the case of ammonia or ammonia water and muriatic acid.

The pairs of electrons marked as dots between two atoms should not be assumed to be located there but as rapidly encircling both atoms and thus appearing more as an extended electron cloud. The other nonbonding electron pairs are localized and occupy certain spherical sectors of these atoms. It is easier to convey this information in models than on paper, especially when a large number of bonds must be described.

Another common way of expressing the electron distribution around the atoms in a molecule utilizes a dash to symbolize each electron pair *(5)*. This is used whether standing for a pair in a shared bond or for an isolated pair. The importance of the latter electron pairs will be understood when the nature of secondary bonds is described.

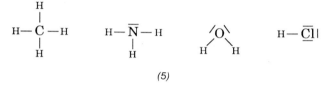

(5)

Since all materials occupy a certain three-dimensional space, it is impossible to fully express their features on a flat piece of paper. In the case of methane, the masses of all four hydrogen atoms are equally distributed around the central carbon atom. Each distance between hydrogen and carbon is 0.109 nm, which is obtained by adding the atom radii of both atoms. Each of the six bond angles H—C—H is 109.5°. It is important to keep these spatial arrangements in mind for any chemical formula expressed on a flat piece of paper.

The corresponding geometrical shapes for these four compounds, methane, ammonia, water, and hydrogen chloride are illustrated in Figure 3.2.

The distances represent the sum of the valence radii established for the various elements forming covalent bonds. The respective valence radii values for some important elements are also given in Table 3.1.

When comparing the valence angles of the three atoms of interest, carbon, nitrogen, and oxygen, only the valence angle of carbon corresponds exactly to an angle of a crystallographic body (see Fig. 3.2). The angle of 109.5° is formed between any of the four corners of a tetrahedron with the carbon atom in their center. The valence angles of nitrogen and oxygen are somewhat diminished as a result of the inequality between the bonding electron pairs and the one or two isolated nonbonding electron pairs. It will be shown later that in many cases the small differences in valence angles can still be accommodated in structures paralleling those structures that would have only carbon atoms in their backbone. Examples are polyoxymethylene, nylon, and cellulose.

All these distances are taken from the center (nucleus) of the respective atoms. Every atom occupies a certain sphere, which is best defined by the van der Waals radii, numerical values of which are also given in Table 3.1. The total volume of the molecule will appear more like the illustrations shown in Figure 3.3.

The four preceding examples represent only the starting points for organic compounds. To build larger molecules one can first obtain the following radi-

		Bond length (nm)	Bond angle H—X—H (°)
Methane	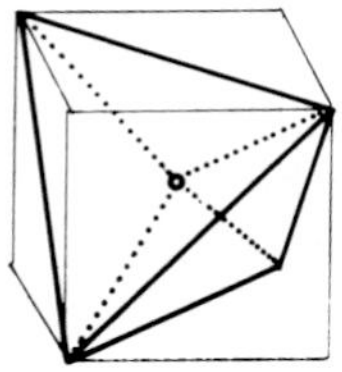 Tetrahedron	0.109	109.5
Ammonia	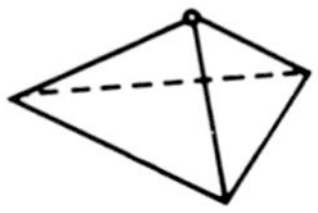 Triangular obtuse pyramid	0.102	107.0
Water	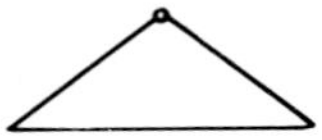 Triangle	0.096	104.5
Hydrogen Chloride	 Pointed rod	0.127	—

FIGURE 3.2.   Bond lengths and angles for a tetrahedron, triangular obtuse pyramid, triangle, and pointed rod.

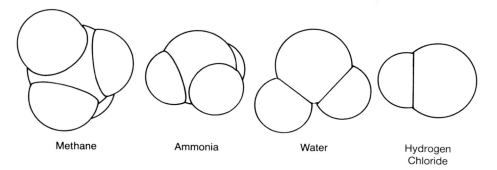

FIGURE 3.3.   Space-filling calotte models for these four molecules.

cals from each of these molecules by removing one hydrogen atom, as shown in (6).

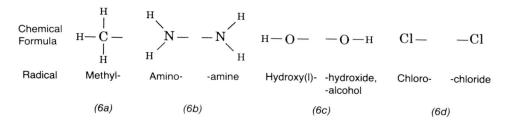

These radicals are materially nonstable, but the following compounds can be derived from them by combining them on paper with another radical, in the following examples a methyl radical. The resulting compounds are shown in (7) and (8):

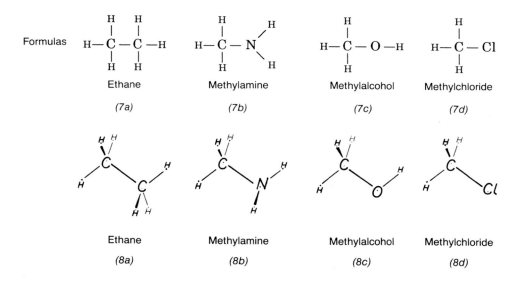

In addition to the "flat" chemical formulas found in most texts, steric three-dimensional structures are also shown to introduce the reader to the space-relating atomic dispositions that will be displayed in this book wherever possible.

A C—C bond should not be considered a stiff connection of these two atoms since both atoms retain their ability to rotate freely around the bond axis. This freedom becomes expressed by the multitude of possible directions of the other three bonds emanating from each of those carbon atoms. Still—as is the case in diamond—the most stable conformation must be visualized as if the directions of each of the three subsequent bonds were displaced by an angle of 180°, forming a staggered trans conformation instead of a cis eclipsed one. The other less favored positions (plus or minus gauche), observed in other crystalline polymers forming spirally arranged conformations, will be described later. Figure 3.4 illustrates this most stable arrangement by comparing the six bonds existing in any one of the chain links in polyethylene crystals with the

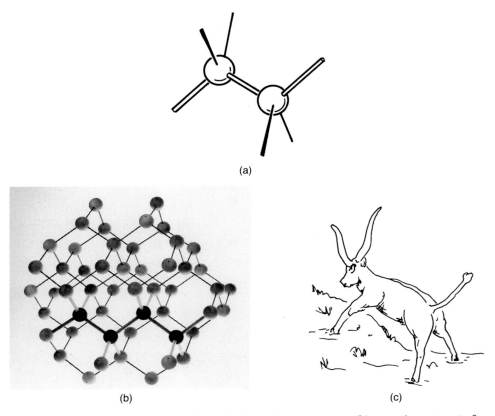

(a)

(b)   (c)

FIGURE 3.4.   (a) Isometric view of an ethylene chain segment, (b) a section cut out of a diamond crystal, and (c) a three-legged bighorn ox, respectively. All three illustrations present the same angular disposition.

extremities of a bighorn ox that spreads its rear legs to compensate for the loss of one of its front legs. (Both the horns and the rear legs are spread 109°.) At the left, the same arrangement of the bonds of a four carbon atom chain portion is shown as it looks when superimposed onto the diamond lattice. Most of the isometric (axoniometric) views of chemical formulas in this book represent the same point of view as sketched in this figure.

In descriptions of more complicated organic molecules the hydrogen atoms are frequently eliminated. This not only improves the visualization of important bonds responsible for the build-up of larger molecules but is also justified since hydrogen atoms, because of their low atomic weight, contribute little to the mass of the polymer. The illustrations in this book—for the most part—compromise by symbolizing hydrogen atoms with a much smaller letter H. The dots, but not the letter itself, indicate the position of the atoms.

If another methyl group substitutes for one of the hydrogen atoms in the above illustrated compounds, the possible formations are shown in (9) and (10).

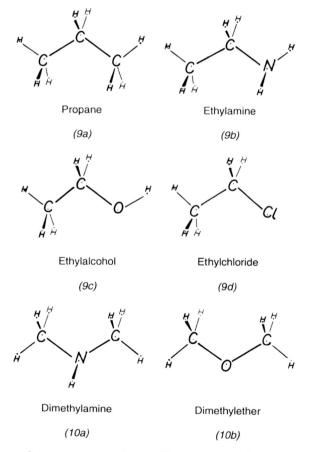

Propane

(9a)

Ethylamine

(9b)

Ethylalcohol

(9c)

Ethylchloride

(9d)

Dimethylamine

(10a)

Dimethylether

(10b)

This build-up of new compounds could be repeated many times and also expanded in many ways by including other kinds of radicals at will.

In all these cases, the relative spatial arrangements of the atoms—bond radii and bond angles—remain practically unchanged. If more than three atoms are combined in one row, compounds with the same chemical composition may occur in different structures and exhibit different properties as a result of isomerism. Details of various isomerism effects will be found in Chapter 4.

The practical reactions employed by the chemical industry for the production of plastic raw materials are quite different. They have changed over the years and are still occasionally modified. Their description is beyond the scope of this text.

The covalent chemical bond is not restricted to single bonds but can also be extended to double or triple bonds. The radicals in *(11)* may be used as examples.

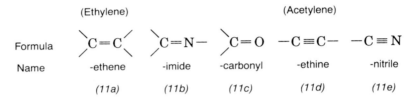

	(Ethylene)			(Acetylene)	
Formula	$\diagup C=C \diagdown$	$\diagup C=N-$	$\diagup C=O$	$-C\equiv C-$	$-C\equiv N$
Name	-ethene	-imide	-carbonyl	-ethine	-nitrile
	(11a)	*(11b)*	*(11c)*	*(11d)*	*(11e)*

Because compounds with double or triple bonds can be converted to saturated, single bond compounds by the addition of two hydrogen atoms, they are called unsaturated compounds.

The covalent double bond connects the two atoms rigidly, restricting a free rotation about the bond. In *n*-butane, a saturated hydrocarbon with four carbon atoms in a row, the three bonds retain their free rotation resulting in an unlimited spatial arrangement of those carbon atoms (see Fig. 4.3 on p. 41). The analogous unsaturated compound occurs only in two very distinct forms exhibiting widely varying properties *(12)*. For both compounds all carbon atoms must be positioned in one plane. This differentiation also becomes important for many polymers, e.g., natural rubber.

Covalent C—C bonds of a number of compounds must be assumed to lie between a single and a double bond since the rule that each carbon atom

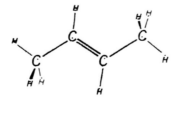

trans-butylene

(12a)

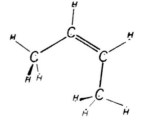

cis-butylene

(12b)

always exhibits four bonds must also be maintained for compounds in which fewer than four atoms surround each carbon atom. Both examples are derived from ring structures having six carbon atoms in one ring.

Benzene (C_6H_6), in which each carbon atom is connected to one hydrogen atom and two carbon atoms, is the first example. The compound's remaining valences are split and each half is added to a neighboring carbon atom. All six carbon atoms are considered equal in all respects since they are equally spaced and rigidly positioned in a plane. The possible expressions for this type of bond are shown in (13).

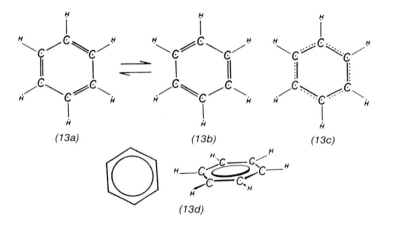

(13a)   (13b)   (13c)

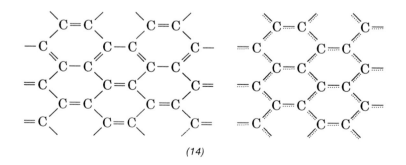

(13d)

Chemical formula for benzene: (13a) and (13b) express the resonance of the double bonds, (13c) expresses equilibrium in the distribution of bonding electrons, and (13d) expresses the truncated formula as commonly used in projected and isometric view.

The arrangement of the carbon atoms in graphite is very similar, with each carbon atom connected to three other carbon atoms. This structure (14) makes it necessary to allocate an additional one-third of a bond to each C—C bond.

(14)

In both substances, the C—C—C angles are exactly 120°. Both represent flat disks and the distances between the carbon atoms lie, as expected, between the distances for single and double carbon bonds (see Tables 3.1 and 3.2).

IONIC BONDS

In addition to the covalent bonds connecting electrically neutral atoms, the
ionic bonds exist between ions, which are electrically charged atoms or groups
of atoms. As seen in the periodic table of elements, and the presentation in
Figure 2.1, atoms on the left-hand side have only one or a few electrons in their
outer shell whereas atoms on the right-hand side have one or only a few
vacancies in their outer electron shell.

The noble gases represent the most stable elements. Their outer electron
shells are fully occupied by eight electrons in the case of argon and neon. The
adjacent elements (just left and right of them) have a great tendency to acquire the same full set of outer electrons.

The metallic elements on the left-hand side are inclined to give up one
electron and thus form a positively charged ion (cation). The halogen elements
tend to acquire an electron and form a negatively charged ion (anion). After
those exchanges have taken place, both kinds of ions will possess a full comple-
ment of electrons like the noble gases but will be electrically charged. Two
examples *(15)* and *(16)* should be noted.

$$\text{Na·} \quad + \quad \text{·}\ddot{\text{Cl}}\text{:} \quad \rightarrow \quad [\text{Na}]^+ \quad + \quad [\text{:}\ddot{\text{Cl}}\text{:}]^-$$

Sodium Atom	· Chlorine Atom	Sodium Cation	Chlorine Anion
(15a)	(15b)	(15c)	(15d)

$$\text{·}\dot{\text{Si}}\text{·} \quad + \quad 2 \; \text{·}\ddot{\text{O}}\text{:} \quad \rightarrow \quad [\text{Si}]^{4+} \quad + \quad 2[\text{:}\ddot{\text{O}}\text{:}]^{2-}$$

Silicon Atom	Oxygen Atom	Silicon Cation	Oxygen Anion
(16a)	(16b)	(16c)	(16d)

Formation of sodium chloride [table salt, melting point 1470°F (800°C)] and silicon dioxide [quartz,
melting point 2658°F (1459°C)].

The ionic bond is the simplest interatomic binding force resulting from the
electrostatic attraction between positively and negatively charged ions. In contrast to the covalent bonds, the ionic bonds do not connect two distinct ions and
do not form molecules but create attractive spheres around each ion that cap-
ture a number of oppositely charged ions. Depending on the size of the ions,
their electrical charge, and the space available, one ion may be equally surrounded by four, six, eight, or so ions of opposite electrical charge. Electrical
neutrality is maintained due to the presence of equal numbers of positive and
negative charges in crystals and solutions.

The high bond strength of ionic bonds is responsible for the high modulus of
elasticity and generally very high melting points of ionically bonded materials.

Many ionic bonds reveal a great weakness for liquids with a high dipole
moment such as water (based on the triangular positioning of the atoms).
These solvent molecules can extricate ions from the surface and thus readily
dissolve many ionic crystalline materials. These ions will gain free independent mobility in the solution, the only restriction being that equal amounts of
oppositely charged ions must be present in all volume entities.

METALLIC BONDS

The *metallic bond* is least understood. In covalently bonded materials the atoms must always be arranged so that well-defined bonds can be identified. In ionically bonded materials the efficient occupation of space becomes more important but the requirement for obtaining an overall electrical neutrality will impose certain restrictions. In the case of metals even this restriction is also eliminated. Therefore in metals the atoms are arranged in a geometrically most-compact way.

Two frequently occurring atomic arrangements are illustrated in Figure 3.5. At left the face-centered cubic and at the right the close-packed hexagonal structures are shown. In the upper row the centers of the atom nuclei are marked, whereas the bottom photos give an indication of the space filling capabilities of both structures. These dimensions should be regarded as corresponding to the van der Waals radii commonly applied to covalently bonded substances. In both cases each metal atom touches twelve neighboring atoms.

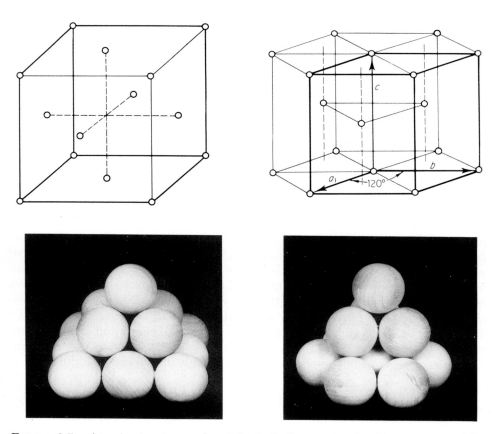

FIGURE 3.5.   Atomic structures of metals. Left: face-centered cubic structure; right: close-packed hexagonal structure. (Drawings reproduced with permission by Glenco Division of the Macmillan/McGraw-Hill School Publishing Co. from Introduction to Physical Metallurgy by S.H. Avner, 1974.)

The distribution of electrons in metallic bonds is not as clear as it is in covalent or ionic bonds. The assumption that at least some of the electrons remain capable of free mobility in the interstices explains best the peculiar optical and electrical properties of metals (light reflectivity and electrical and thermal conductivity).

The main weakness of the metallic bond is founded on the imbalance of the bonding electrons at the surface, which makes most metals so vulnerable to chemical reactions leading to the well-known corrosion processes.

Only very few covalently bonded compounds (e.g., members of highly cross-linked resins such as amino or phenolic resins) have their primary bonds spread out sufficiently so that structural parts could be made out of them. In most cases—and this includes all thermoplastic polymers—these entities are too small, meaning that additional binding forces must exist that are capable of cementing several molecules together. Only then can these agglomerates be put to use to form structural parts.

SECONDARY BONDS

A great variety of *secondary bonds* exists. They are capable of turning an otherwise gaseous or liquid material of no strength into a strong solid. For all practically used structural polymeric materials, the balanced combination of covalent bonds and these secondary bonds is a prerequisite.

van der Waals Bonds

It has been assumed that the *van der Waals force* attracts all atoms universally in a nondirectional way similar to the gravitational force. This force, causing inert gases to be converted to liquids and solids at very low temperatures, is considerably weaker than any one of the other primary bonds described. Therefore, the proximity to which atoms of neighboring molecules will approach each other is much greater than the covalent valence bonds. Both important distances and radii can be seen in Tables 3.1 and 3.2. If molecules are pictured expressing the van der Waals forces, the spatial expansion of those molecules becomes clarified, giving a good indication of how several molecules might be best arranged to efficiently fill a certain space. Doubtless the expression of all chemical formulas only as calotte models is, however, not practical. The atomic ball-and-stick model more likely shows all bonds and all atoms in a picture, whereas the space-filling model will sometimes hide some of them. Examples are given in the photos of the molecule of styrene in Figure 3.6.

Hydrogen Bonds

The *hydrogen bond* ranks next in importance among secondary type bonds for plastic materials. Although its bond energy is only about one-tenth the energy of covalent bonds, its contribution to increased mechanical strength and

FIGURE 3.6.   Picture of styrene molecule. (a) Ball-and-stick model; (b) space-filling, spherical calotte model.

higher melt temperatures can become significant as illustrated for two representative plastics:

	Tensile strength		Melting temperature	
	psi	MPa	°F	°C
Polyethylene, high density	4000	25	275	135
Nylon polyamide	10,000	70	420	215

Hydrogen bonds can be established between hydrogen atoms bonded to an oxygen or nitrogen atom and other spatially closely positioned atoms having at least one nonbonding electron pair. Examples are the elements nitrogen, oxygen, and fluorine. The great importance of hydrogen bonds is documented by their power to combine the small water molecules to form a liquid and to structure the highly flexible polypeptide chains into distinct helixes in living organisms. A graphic example is given on p. 304 relating to collagen.

The atomic arrangement of typical hydrogen bonds involving oxygen or nitrogen is given in (17). The noted distances reflect the appropriate bond strength of the respective bonds.

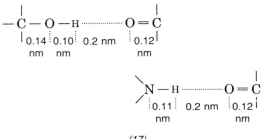

(17)

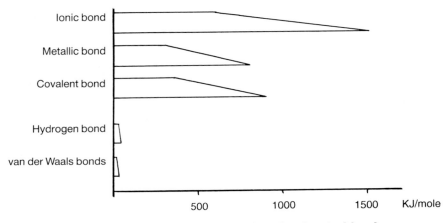

FIGURE 3.7.   Bond strength values for chemical bonds.

Dipole Moments

Other forces acting on molecules are caused by the electrical imbalance of some atoms or atomic groups. They are called *induced* or *permanent dipole moment forces*. The retention of the high strength of polyvinyl chloride in spite of the addition of large amounts of liquid plasticizers should be cited as an example.

Summary

As already indicated, the bond strength values for the various types of bonds are quite different. In many instances their character cannot be unequivocally resolved and may overlap. The ranges in strength expressed in energy terms are given in Figure 3.7.

REFERENCES

R.M. Brick, A.W. Pense, and R.B. Gordon, Structure and Properties of Engineering Materials, McGraw-Hill, New York, 1977.
W.D. Callister, Jr., Materials Science and Engineering. John Wiley, New York, 1985.
L. Pauling, The Nature of the Chemical Bond and the Structure of Molecules and Crystals. Cornell University Press, Ithaca, NY, 1960.

4

General Concepts Relating to High-Molecular-Weight Plastics

EFFECT OF INCREASING MOLECULAR WEIGHT ON PROPERTIES OF HYDROCARBONS

The properties of any material are based on

1. the atoms or groups of atoms that constitute the molecule,
2. the bonds or structural arrangement of these atoms, and
3. the extent of the aggregation, in the case of plastics materials the size of the polymer.

Throughout this book the endeavor is to combine—as much as possible—the expression of the structural arrangement of all atoms in a molecule with the presentation of its chemical formula. The strong relationship between properties of all structural materials and the arrangement of atoms in those molecules is a fact. Unfortunately only crystalline or oriented structural materials can be properly presented on flat paper. To enhance the three-dimensional perception of the arrangement of atoms in space, the chemical formulas are consistently presented in isometric view and some, as in Appendix C, in stereoscopic pictures. So as not to disturb this perception, idealized angles and geometric forms are used instead of representing the actual slight variations in bond angles and bond distances. This is exemplified by the cellulose molecule, in which none of the six angles in the ring structure are exactly the same. Atoms or groups of atoms that stick out of the plane are shown in **boldface** type and naturally their distances on paper are appropriately shortened. To remind the reader that the hydrogen atom has less than one-tenth the mass of the other atoms illustrated, it is shown in smaller type (a small H) or—as frequently seen—is omitted. Smaller molecules, such as monomers, are drawn as they would appear in stretched-out polymers or in crystals even though this seldom holds true.

Linear Arrangements of Carbon Atoms

To observe the changes that will occur if the size of a molecule is increased and its composition, atomic constituents, and bond relationships remain the same, a study of these effects should start with the simplest organic molecule, methane (*1a*). It consists of one carbon atom and four hydrogen atoms. Combining two carbon atoms will leave six valences, which can be occupied by six hydrogen atoms. Once occupied, the molecule ethane (*1b*) is formed, which has a

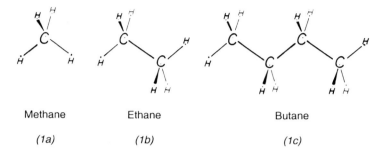

Methane	Ethane	Butane
(1a)	*(1b)*	*(1c)*

molecular weight of not quite twice that of methane (two hydrogen atoms have been eliminated). Although both molecules are gases at ambient temperatures their melting points, boiling points, and specific gravities are quite different. When the number of carbon atoms in the molecule is again doubled, a similar change in properties is noted. This increase in the number of carbon atoms can be repeated several times. However, as molecular weight further increases, the magnitude of the changes in properties diminishes more and more. Some of these compounds have been selected in Table 4.1 and their composition and properties listed. All consist of a certain number of chain links —CH_2— (methylene) and of two end groups —CH_3 (methyl).

All members containing up to four of these chain links are gases, those having up to 15 are liquids, and those with more are crystalline solids at room temperature.

As molecular weight increases, the carbon content also increases, from 75.0 to 85.7%. Decreasing hydrogen content is responsible for a decreasing heat of combustion and ease of flammability. Specific gravity can be compared only between the liquid and solid members of this group, with a limiting value at about 0.95 g/cm³. The boiling points increase rapidly at quite proportional intervals as molecular weights increase. The high-molecular-weight alkanes will thermally decompose prior to volatilization. Their melting points also show a general increase up to a limit of 275°F (135°C). Because of the zigzag order of the chain links, greater differences may occur in the crystal order of the shorter chain alkanes—especially at lower molecular weights. This results in minor melting point discrepancies with the odd-numbered alkanes having lower melting points.

When comparing the last two entries of Table 4.1, it is important to note that high-density polyethylene and the ultrahigh-molecular-weight polyethylene exhibit no changes in the properties listed with the exception of molecular weight.

TABLE 4.1. Properties of Linear Straight Chain, Saturated Hydrocarbons

Name	Number of $-CH_2-$ Chain Links	Molecular Weight	Carbon Content (%)	State of matter	Specific Gravity (g/cm³)	Melting point		Boiling point	
						°F	°C	°F	°C
Methane	1	16	75.0	Gas	0.000,72	−299	−184	−263	−164
Ethane	2	30	80.0	Gas	0.001,36	−278	−172	−127	−88.5
Propane	3	44	81.8	Gas	0.002,02	−310	−190	−48.1	−44.5
Butane	4	58	82.8	Gas	0.002,7	−211	−135	33	0.6
Pentane	5	72	83.3	Liquid	0.633,7	−204	−131	97	36.2
Hexane	6	86	83.7	Liquid	0.660,3	−138	−94.3	156	69.0
Octane	8	114	84.2	Liquid	0.704,2	−70	−56.5	259	126
Hexadecane	16	226	85.0	Solid	0.771	68	20	518	270
Paraffin	~30	~500	85.3	Solid	0.86–0.93	108–167	42–75	—	—
High-density linear polyethylene	~9000	~0.125 × 10⁶	85.7	Solid	0.941–0.965	259–277	126–136	—	—
Ultrahigh-molecular-weight polyethylene	>100,000	>1.5 × 10⁶	85.71	Solid	0.945	275	135	—	—

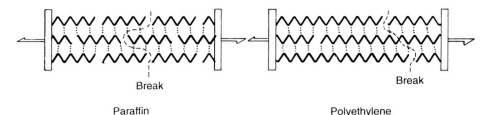

Break   Break

Paraffin   Polyethylene

FIGURE 4.1. Schematic representation of van der Waals forces acting on paraffin and polyethylene. Dotted lines represent van der Waals bonds. Dashed lines indicate where breaks will most likely occur.

All materials contained in Table 4.1 represent straight chain alkane molecules having the same C—C bond strength. This includes the paraffin chain and the polyethylene chain. The tensile strength experienced is nearly zero for paraffin since the great number of stacked or staggered short chains in the bulk material may easily slip by each other if subjected to a shear force. In the case of polyethylene the same low van der Waals forces apply. However, since the polyethylene molecule is considerably longer, the greatly multiplied number of these weak bonds will grab the adjacent chain—sometimes indirectly—at many points, thus preventing slippage. The sum of these weak bonds can become stronger than the strong C—C bond within the molecule, resulting in a C—C bond failure rather than a slippage lengthwise along the long chain (Fig. 4.1).

However, it should not be assumed that the theoretical C—C bond strength could be obtained in a bulk sample. The reason for such a great discrepancy is that most of the C—C bonds are not aligned in the proper direction and that other imperfections (such as impurities, notches, voids) are widely distributed in all bulk materials. Exceptions to this are carefully prepared materials such as whiskers and maybe some gel-spun fibers.

All compounds discussed to this point represent linear or straight chain hydrocarbons or so called n-alkanes. This designation must already be considered improper since the carbon atoms are arranged along a straight line only in ethane. The next heavier molecule, propane, consists of three carbon atoms, which cannot lie on a straight line but must be arranged in a plane where two lines meet at a 109.5° angle. Since the C—C single bond imparts free rotation to the carbon atoms, the possible variations in the relative positions of the third carbon atom lie somewhere on the circle drawn in Figure 4.2.

By adding a fourth carbon atom as found in butane, the possible shapes this molecule could assume increase considerably. These are sketched in Figure 4.3, and as models photographed in Figure 4.7. The "straightest" arrangement possible occurs only in crystalline butane, in which all four carbon atoms are located in zigzag fashion on one plane. In the liquid and gas phases, the possible positions of the end carbon atoms should be visualized as being projected anywhere on the dotted lines (circles). The dashed lines mark the extreme positions for those bonds. The longest end-to-end distance occurs only in the crystalline state.

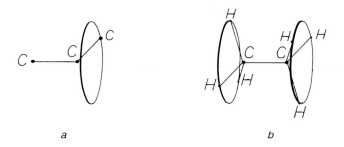

FIGURE 4.2.   Possible positioning of (a) the third carbon atom in propane and (b) the hydrogen atoms in ethane. (C—C axis viewed offset by 15°.)

It is important to visualize all C—C chain bonds as positioned randomly kinked in three-dimensional space with only the distance and the angle—but not the direction—kept constant. The orderly zigzag chain arrangement is restricted to some crystallites, the smallest crystalline domains of solid hydrocarbons. In all organic solid materials—excluding thermoset resins, the free motion of the molecules is impeded by overwhelming clustering of van der Waals forces, but the molecules are strictly ordered only in the crystalline state. Both solvents and increasing temperature can reverse these orders.

The angular positioning is not reserved to carbon atoms but applies to all other atoms or groups of atoms linked to any carbon atom, with the exception of the carbon–carbon triple bond, —C≡C—. The possible spatial placements for the selected four atoms H—C—C—H in ethane are the same as those for the four carbon atoms in butane.

Aliphatic hydrocarbons are not necessarily limited to compounds in which a carbon atom can be connected to only one or two other carbon atoms. Since carbon has a valence of four, possibly all four may be directed toward other carbon atoms, although arrangements with two and three carbon atoms connected to one carbon atom are more prevalent. The simplest example of a *branched hydrocarbon* is isobutane *(2b)*, which is the isomer of normal butane *(2a)*.

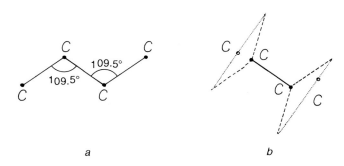

FIGURE 4.3.   Positions of carbon atoms in (a) solid and (b) liquid butane. (All C—C bonds marked with solid lines are positioned parallel to paper plane.)

TABLE 4.2. Properties of Linear Branched, Saturated Hydrocarbons

Name	Number of —CH$_2$— Chain Links	Molecular Weight	Carbon Content (%)	State of Matter	Specific Gravity (g/cm^3)	Melting point		Boiling point	
						°F	°C	°F	°C
n-Butane	4	58	82.8	Gas	0.002,70	−211	−135	33	0.6
Isobutane	4	58	82.8	Gas	0.002,67	−229	−145	14	−10
n-Octane	8	114	84.2	Liquid	0.704,2	−69.7	−56.5	259	126
Diisobutyl	8	114	84.2	Liquid	0.700,0	−132	−91.3	228	109
High-density linear polyethylene	~9000	~0.125 × 10^6	85.7	Solid	0.941–0.965	259–277	126–136	—	—
Low-density branched polyethylene	~20,000	0.3 × 10^6	85.7	Solid	0.917–0.932	204–239	98–115	—	—

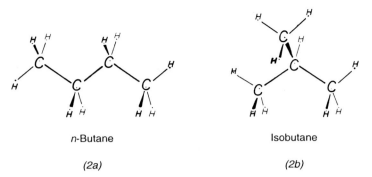

n-Butane   Isobutane

(2a)   (2b)

In Table 4.2 the properties of linear and branched saturated hydrocarbons (alkanes) are juxtaposed and representatives of different molecular weights are listed analogous to the data in Table 4.1. It becomes obvious that all else being equal, a compound with one or a large number of branches (or side chains) must have a lower surface-to-volume ratio and must be more clustered. Also, the molecule's end-to-end distance must be shorter. These changes are responsible for the following property alterations:

1. lower boiling and melting points,
2. less orderly packing, resulting in lower specific gravity,
3. less tendency to crystallize, as well as
4. lower tensile strength and stiffness.

CYCLIC ARRANGEMENTS FOR CHAIN LINKS CONTAINING CARBON ATOMS

There are two additional arrangements of carbon-to-carbon bonds possible in hydrocarbons. As seen in Figure 4.4, ends of carbon–carbon chains with a length of five or six carbon atoms can easily form *ring structures*. The tetrahedron bond angle of 109.5° must be somewhat compressed to 108° to accommodate the five carbon atoms in a regular pentagon. The cyclopentane ring will therefore be under some strain. In a regular hexagon, the angle would be 120°, which is too great to be extended from 109.5°. However, a six-carbon atom ring with a 109.5° bond angle can be obtained free of constraint by kinking a flat hexagon twice at opposite carbon atoms by 125.5°. The two steric configurations of cyclohexane's "boat" and "chair" form are shown in Figure 4.4. A more detailed description is given in Chapter 13.

Before the other possible arrangement for six carbon atoms in the shape of a hexagon is discussed, the *unsaturated carbon bond* must be described. In addition to ethane, two carbon atoms may form another hydrocarbon, ethylene, which contains only four hydrogen atoms as shown in Figure 4.5. To maintain the four valences for each carbon atom, ethylene must bear a carbon–carbon double bond. To uphold the tetrahedron bond angle of 109.5°, ethylene could be drawn as though two tetrahedrons, which have carbon atoms at their centers, join along an edge (two points) and not at one point as in ethane. If this configuration is drawn flat with all atoms in one plane, the resulting angles

TABLE 4.3. Properties of Cyclic Hydrocarbons

Name	Number of —CH$_2$— Chain Links	Molecular Weight	Carbon Content (%)	State of Matter	Specific Gravity (g/cm^3)	Melting point		Boiling point	
						°F	°C	°F	°C
n-Hexane	6	86	83.7	Liquid	0.6603	−138	−94.3	156	69.0
Cyclohexane	6	84	85.7	Liquid	0.7791	43.5	6.4	177	80.8
Benzene	(6)	78	92.3	Liquid	0.8786	42	5.5	176	80.1

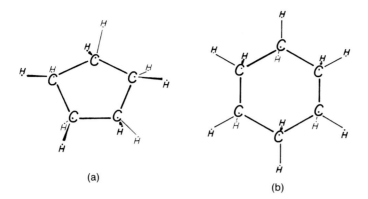

(a)

(b)

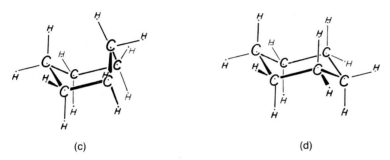

(c)

(d)

FIGURE 4.4.   Angular configuration of saturated hydrocarbon ring structures. Chemical formulas for cyclopentane (a) and cyclohexane (b, c, and d) where in a planar projection (b) all 109.5° angles appear to be 120°, three-dimensional views distinguish between the "boat" (c) and "chair" (d) form, which correspond to the cis and trans configuration.

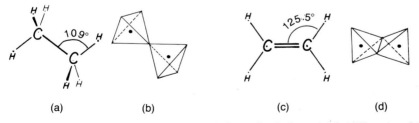

(a)   (b)   (c)   (d)

FIGURE 4.5.   Atomic structures for ethane (left) and ethylene (right) illustrated in two different ways. (•) Position of carbon atoms. (a) For ethane there are never more than four atoms on one plane. (b) The four valences of the carbon atoms are positioned at tetrahedron corners. (c) For ethylene all atoms are on one plane. (d) The two valences of the carbon atom double bond are positioned at the tetrahedron edge.

are 109.5° between the H—C—H bonds and 125.5° between the C=C—H bonds.

By transposing these relationships to the planar six-carbon atom hexagon arrangement, the benzene molecule will be obtained. It contains six carbon and six hydrogen atoms. Under these conditions all valences can be satisfied only if it is assumed that the carbon atoms are alternately bonded by C—C single and double bonds, or all bonds are quasi one and a half C—C bonds. The angular corrections required are only 5.5° to obtain the called for 120° angle. Since there is only one valence left per carbon atom, all six hydrogen atoms must be positioned in the same plane as the carbon atoms. There is more than one possible way of writing the structural formula for benzene that will explain all its properties and its chemical reactions (see formulas on p. 31).

To complete the information given in Tables 4.1 and 4.2, the same properties are listed in Table 4.3 to illustrate the effects of ring formations. The physical properties of the two cyclic hydrocarbons are compared with the *n*-hexane molecules. The sharply increased high melting points of the two cyclic hydrocarbons attest to the nearly equally high rigidity of those two rings. These effects reappear in polymers when aliphatic chain links are substituted by these six-ring structures. (See also high-heat polycarbonate, p. 112.)

Three-Dimensional, Steric Arrangements of Carbon Atoms

As the reader is now familiar with the various C—C bonds found in organic materials, the structures of diamond, lonsdaleite, and graphite should be recalled. Though crystallographers correctly classify diamond as a cubic crystal with eight carbon atoms positioned in the corners of a cube (Fig. 4.6), the arrangement of the other carbon atoms lacks clarity. Four of the carbon corner atoms are dangling in space with no connections to any of the other carbon atoms shown, as recognizable from the presentation on the right-hand side, where the thin lines marking the unit cell have been omitted. This presentation appears more like five-petal flowers drawn on square checkered paper since none of the carbon bonds has a simple relationship to any of the lines of the lattice grid. Therefore—as predominantly seen in the literature (likewise in Fig. 4.7 and Fig. 2.4, p. 9), the diamond structure is also represented as belonging to the hexagonal system though its symmetry is much higher, because dimensions in all four directions are identical. In this fashion the relationship between the structure of diamond and the hexagonal unit cells of the other two carbon modifications, lonsdaleite and graphite, becomes very clear.

If graphite is seen as a laminar structure in which each carbon atom participates in three benzene-ring structures, then each carbon atom in diamond participates in 12 cyclohexane-ring "chair" structures. It also becomes clear that lonsdaleite shows certain features of both of these structures. Of the 12 cyclohexane-ring formations, only three represent the "chair" form; the other nine represent the "boat" form. Figure 4.8b illustrates the lonsdaleite crystal, showing the carbon atoms with all their chemical bonds. The high strength of lonsdaleite and diamond is based on the total presence of strong bonds toward all four adjacent carbon atoms.

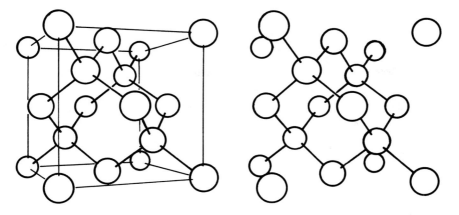

FIGURE 4.6.   Unit cell of diamond. From Ivan Bernal, Walter C. Hamilton, and John S. Ricci, Symmetry. Copyright © 1972 by W.H. Freeman and Company. Reprinted with permission.

As previously indicated, the structural arrangement of carbon atoms found in diamond or graphite can be used to illustrate the arrangements of carbon atoms in organic and polymeric materials. This is especially true for the ordered arrangements found in crystalline materials.

A few examples of how organic molecules, including sections of macromolecules, can be superimposed onto the diamond lattice are given. The left image (Fig. 4.7a) shows the alignment of the locked-in C—C sections in the crystalline butane. In all these figures the C—C bonds are visible as black lines, the C—H bonds at 0.11 nm are shorter than the bonds in diamond (0.15 nm), but the very weak van der Waals bonds are much wider (about 0.24 nm). To the right (Fig. 4.7b), within the outlines of diamond one, the image shows a possible, but very improbable arrangement of the C—C bonds for liquid butane have shown. As a result of the free rotation at each successive bond, a great variety of positions exist. Below (Fig. 4.7c), the carbon atoms forming cyclohexane rings (only "chair" form) are highlighted.

As has been shown, a material may be made stronger by increasing the number of strong C—C bonds within a unit of volume. This can be accomplished by extending the C—C chain of crystalline butane along the zigzag line of the diamond lattice to obtain the strong crystalline polyethylene, also shown in Figure 6.2 (p. 138) and Figure 6.6 (p. 147).

The relationship between lonsdaleite and graphite becomes clearer when all the carbon atoms in the horizontal planes of the cyclohexane "chair"-type ring structures are merged into a single plane forming carbon hexagon rings as shown in Figure 4.8a. With this arrangement the distance of the C—C bonds is reduced from 0.154 to 0.142 nm.

Since the specific gravity of graphite is lower (2.26) than that of diamond (3.51), the distance between adjacent planes of these strung-up hexagons must emerge considerably extended (0.339 nm). The benzene ring also consists of a planar hexagon-shaped ring (see p. 31). All the C—C—C angles in both rings

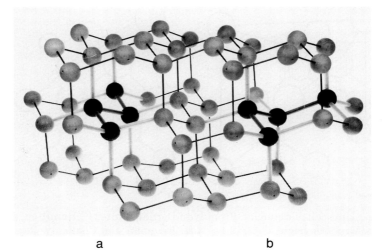

a   b

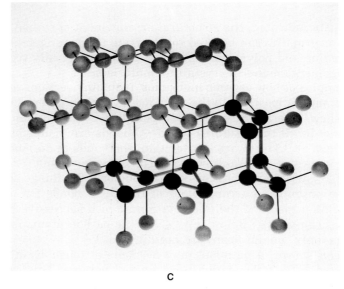

c

FIGURE 4.7.   Photos show how aliphatic compounds can be positioned onto the diamond lattice. (a) Crystalline butane, (b) liquid butane, (c) cyclohexane.

must be the same (120°), but there is a very small difference (0.002 nm) between the C—C bond length in graphite (0.142 nm) and in benzene (0.140 nm). The significance of the benzene ring is that those six carbon atoms are very rigidly held together and have a much greater inertia than six carbon atoms lined up in an aliphatic chain. This is a trait shared by all aromatic compounds and is expressed in the higher modulus of elasticity and diminished creep behavior of aromatic ring-containing polymers. The cyclohexane ring struc-

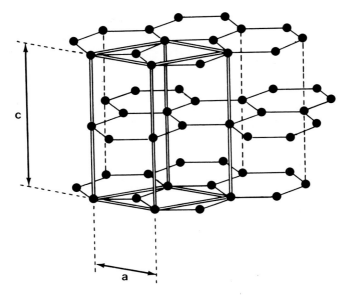

a

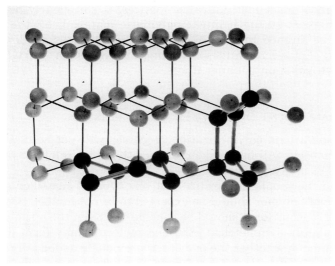

b

FIGURE 4.8.   Comparison of the carbon reticulation of (a) graphite and (b) lonsdaleite. In lonsdaleite both the "boat" and "chair" form of the cyclohexane ring are highlighted. Graphite structure from D.J. Johnson, High-temperature stable and high-performance fibres. F. Happey, ed., Applied Fibre Science, Vol. 3 Academic Press, New York, 1979.

TABLE 4.4.   Effect of Intensification of C—C Bond Clustering
in Materials on Modulus of Elasticity

Material	Modulus of elasticity	
	10^6 psi	MPa
Polyethylene	0.03–0.15	200–1000
Polycarbonate	0.35	2,500
Polystyrene	0.45	3,000
Polyethylene terephthalate	0.5	3,500
Phenolic resins	1.0	7,000
Wholly aromatic polyesters	2.0	14,000
Nearly perfectly oriented polyethylene fibers	15	100,000
Graphite fibers	75	500,000
Diamond and lonsdaleite	163	1,120,000

ture (also similarly shared by cellulose compounds) approaches the rigidity of
the benzene ring.

Table 4.4 illustrates how the concentration and the orientation of strong zigzag C—C bonds affect mechanical properties. It can clearly be seen that to
obtain the highest strength linear polymeric material, all the strong C—C
bonds must be aligned in the direction of applied stress. Under these conditions, moduli of elasticity values exceeding 15 million psi (100 GPa) have been
observed with gel-spun polymer fibers.

Steric Arrangement of Polymer Chain Links

Although the various polymerization processes will not be discussed in this
book, some aspects will be considered since they also relate to properties and
applications of many plastics.

Most basic thermoplastics utilize just one type of monomer and therefore
just one type of polymer should be expected to be obtainable. Since this is not
the case, the reasons for it should be considered. Even the structures obtained
from a completely symmetrical ethylene molecule may form different polymers depending on whether the end of the previously added ethylene molecule
accepts the following ethylene molecule or whether it will be added further
back along the chain, resulting in the formation of side branches. Less symmetric monomers such as propylene or vinyl chloride may, in addition, lead to
different polymer structures depending on whether the head or tail ends are
always added regularly or irregularly (3). Another aberration may develop
when the next monomer is added—though consistently head to tail end—
irregularly from either side. These polymers have been called atactic (4). If the
monomer is regularly attached they are termed isotactic (5, 6) (if all asymmetric carbon atoms become right d- or all left l-isomers) and if the monomer
is added alternatively from each side they are termed syndiotactic (7). The
corresponding formulas in the case of polypropylene are as follows:

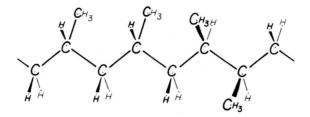

Head-tail Irregularity

(3)

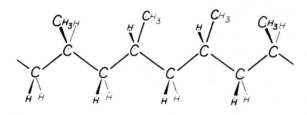

Atactic Irregularity

(4)

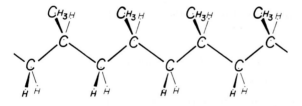

Isotactic Regularity

(5)

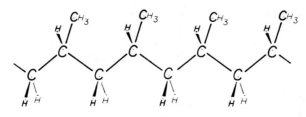

Isotactic Regularity

(6)

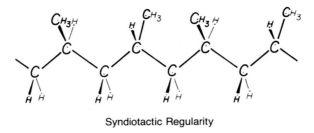

Syndiotactic Regularity

(7)

These stereochemical arrangements represent atomic *configurations* that cannot be altered just by bond rotations. It would be necessary to break the polymer chain to convert one form into another. On the other hand, *conformation* denotes only the geometric arrangement of atoms in the chain, utilizing the principle of free rotation about chemical bonds.

The steric regular varieties have been made available through the utilization of special catalyst systems. Their importance is based on the polymer's ability to be arranged within a crystal lattice whereas the irregularly configured polymer chains are not able to do that.

In most cases the user is not aware of the steric arrangement of the monomer units since plastics are not sold according to either their molecular weight or their steric structure but according to their contemplated processing method or their intended use. The quite rigid polypropylene plastics that border the group of engineering thermoplastics all belong to the steric highly ordered varieties, whereas the irregularly structured, amorphous polypropylene resins are too flimsy and therefore are usually applied for elastomer gum stocks.

PROPERTIES CONFERRED BY FUNCTIONAL GROUPS TO VARIOUS CLASSES OF POLYMERS

In the preceding section it has been shown how increasing the molecular weight of a substance will change its properties without really changing its chemical composition. Now the changes in properties founded on the peculiarity based on the chemical behavior of functional atoms or groups of atoms should be examined.

First a profile should be drawn of some of the important monomers indicating some of the possible ways that they react. To familiarize the reader with names contained in many polymers, their roots to simple organic compounds are shown in Table 4.5. In the first row the basic molecules and their formulas are listed. Subsequent rows contain the chemical formulas of the corresponding radicals.

Radicals are groups of atoms that recur in many compounds and represent a certain portion of a generally small molecule *(e.g., 8a)*. They contain one free bond. Biradicals *(9b, 10b)*, with two free bonds, can combine with many such radicals by forming polymers. Radicals are often abbreviated by R, or R′ if chemically different. In some instances the word radical is also used to desig-

TABLE 4.5. Basic Chemical Compounds and Their Radicals

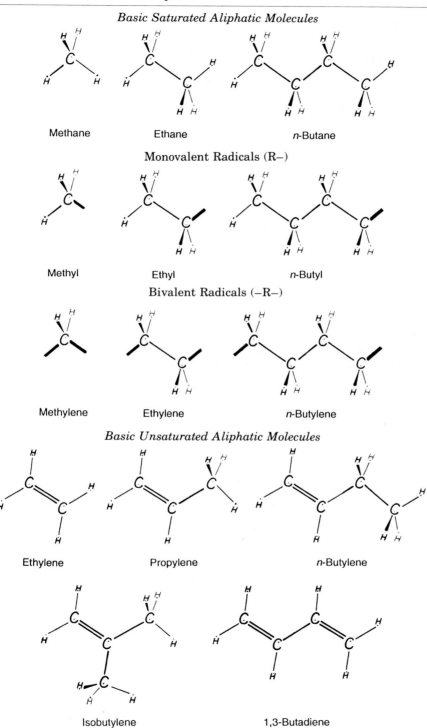

Basic Saturated Aliphatic Molecules

Methane   Ethane   *n*-Butane

Monovalent Radicals (R–)

Methyl   Ethyl   *n*-Butyl

Bivalent Radicals (–R–)

Methylene   Ethylene   *n*-Butylene

Basic Unsaturated Aliphatic Molecules

Ethylene   Propylene   *n*-Butylene

Isobutylene   1,3-Butadiene

(*continued*)

TABLE 4.5. (*Continued*)

Monovalent Radical (R–)

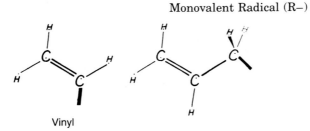

Vinyl

Bivalent Radical (–R–)

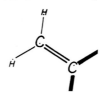

Vinylidene

Basic Cyclic or Aromatic Molecules

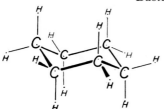

Cyclohexane

Benzene   Naphthalene   Biphenyl, Diphenyl

Monovalent Radical (R–)

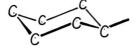

Cyclohexyl

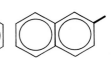

Phenyl   α-Naphthyl   β-Naphthyl

Bivalent Radical (–R–)

Cyclohexylene

m-Phenylene
1,3-phenylene

o-Phenylene
1,2-phenylene

p-Phenylene
1,4-phenylene

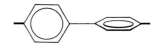

4,4′-Biphenylene

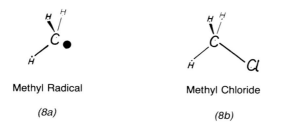

Methyl Radical   Methyl Chloride

(8a)   (8b)

nate a very reactive species of compounds in which a chemical bond is assumed to be broken. Their chemical formulas are characterized by a dot, signifying that only one electron is left with the radical instead of a pair of electrons that belong to a chemical bond, denoted by a dash. Although radicals of monomers usually form after reacting with polymerization catalyst radicals, this relationship has been omitted here for the sake of simplicity.

In cases in which one compound bears the same name as a radical (e.g., ethylene) notice that both have identical numbers and positions of atoms. The difference lies only in the local assignment of the bonding electrons.

Table 4.6 lists a number of monomers that are all characterized by the presence of an unsaturated vinyl group bearing the C=C double bond. They are all quite reactive and many must be stabilized to prevent autopolymerization. Their conversion into the polymer, bearing the same name plus the prefix poly-, via a polymerization reaction is called addition polymerization. It may be illustrated in a simplified way (9a–c, 10a–c, 11a–c).

Formation of Polymer Chains

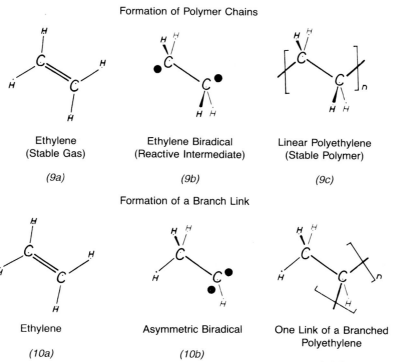

Ethylene   Ethylene Biradical   Linear Polyethylene
(Stable Gas)   (Reactive Intermediate)   (Stable Polymer)

(9a)   (9b)   (9c)

Formation of a Branch Link

Ethylene   Asymmetric Biradical   One Link of a Branched
                                                    Polyethylene

(10a)   (10b)

(10c)

Formation of Diene Polymer Chains

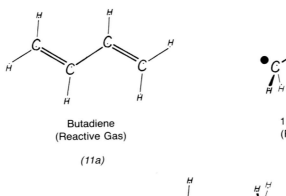

Butadiene
(Reactive Gas)

(11a)

1,4-Butadienyl Biradical
(Reactive Intermediate)

(11b)

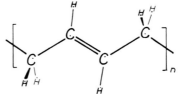

Poly-1,4-butadiene
(Polymeric Elastomer)

(11c)

The ends of linear polymers must be capped by end groups during the termination of chain growth. Since the weight fraction of those end links in a high-molecular-weight polymer is rather low, their contribution to the polymer's physical properties can, in most cases, be disregarded. The end groups, however, may significantly affect the chemical stability of the polymer (unzipping of chain links). More importance must be attributed to the effect of the presence of the loose ends connected to an otherwise restrained chain. The property changes observed on polymers of different molecular weight reflect on their concentration.

The structure of the backbone of all the polymers prepared from the monomers shown in Table 4.6 will basically have an identical carbon–carbon sequence along the whole polymer chain. Therefore, the great differences in properties of these plastics must be attributed to the pending atoms or functional groups that replace one or more hydrogen atoms of the monomer.

All the polymers consisting only of *aliphatic hydrocarbons* are characterized by excellent moisture resistance, low water absorption, and low solubility in organic solvents. They are also resistant to nonoxidizing acids and bases. Their stiffness depends a great deal on their readiness to crystallize, which in turn is influenced by the way they have been polymerized. Their upper use temperature is quite low, the result of their low crystalline melting points and their sensitivity toward oxidative degradation.

Polystyrene, the vinyl polymer that has *aromatic rings* attached at the side, is quite different, although it also consists of only carbon and hydrogen atoms.

TABLE 4.6.   Monomers Belonging to the Class of Vinyl Compounds

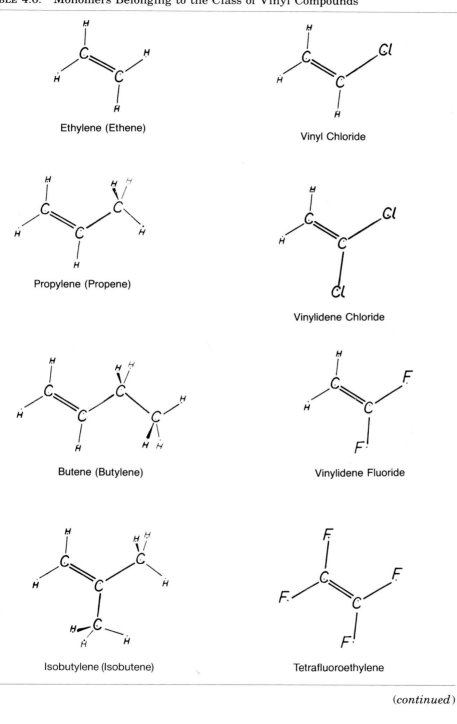

Ethylene (Ethene)

Vinyl Chloride

Propylene (Propene)

Vinylidene Chloride

Butene (Butylene)

Vinylidene Fluoride

Isobutylene (Isobutene)

Tetrafluoroethylene

(*continued*)

Table 4.6.   (*Continued*)

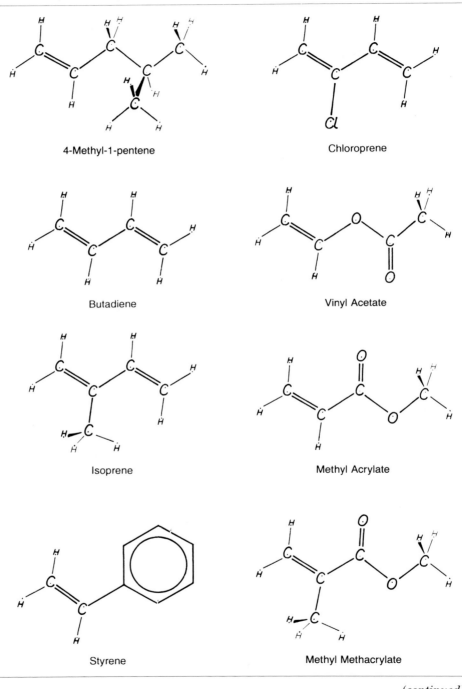

4-Methyl-1-pentene

Chloroprene

Butadiene

Vinyl Acetate

Isoprene

Methyl Acrylate

Styrene

Methyl Methacrylate

(*continued*)

Table 4.6.   (*Continued*)

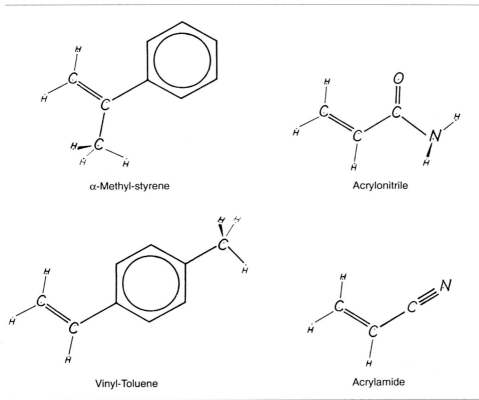

α-Methyl-styrene

Acrylonitrile

Vinyl-Toluene

Acrylamide

The six-membered rings lend stiffness to the otherwise flexible chain. Polystyrene has a modulus of elasticity in the same range as the best engineering plastics; however, because of its severe brittleness it must be copolymerized or blended with other tough resins to become suitable for such demanding applications. These measures also help overcome its solvent sensitivity.

Halogen-containing compounds can be visualized as consisting of alkyl (aliphatic hydrocarbon) radicals that are combined with either chlorine, fluorine, or bromine. The presence of these halogen elements leads to flame-retardant polymers. Their flammability can be further reduced by compounding them with inorganic flame-retardant additives.

Some of the important compounds of this group should be mentioned. *Methyl chloride (12)* is a gas that is sometimes used as a foam-blowing agent. *Methylene chloride (13)* is an important volatile solvent and also a polyurethane blowing agent. *Vinyl chloride (14)* and *vinylidene chloride (15)* are two monomers that impart good solvent, oil, chemical, and weather resistance to their polymers. *Chloroprene (16)* is a monomer leading to oil- and oxidation-resistant elastomers. Its structure resembles that of isoprene.

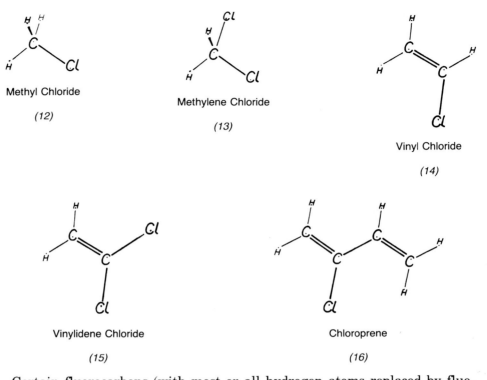

Methyl Chloride

(12)

Methylene Chloride

(13)

Vinyl Chloride

(14)

Vinylidene Chloride

(15)

Chloroprene

(16)

Certain fluorocarbons (with most or all hydrogen atoms replaced by fluorine) represent monomers that result in high temperature, solvent resistant, nonflammable polyfluorocarbon polymer plastics and elastomers. Chemical formulas are shown on p. 102.

Chlorofluoro(hydro)carbons, low-molecular-weight chlorine and fluorine containing gases or liquids, are very useful refrigerants, foaming agents, and solvents, but will have to be phased out or be limited in their use because of their effect on the stratosphere. *Tetrabromo-bisphenol A (17)* is a coreactant contributing self-extinguishing properties to polyesters and epoxies.

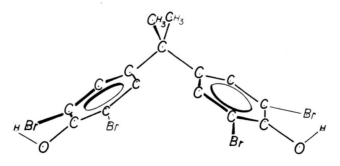

Tetrabromo-bisphenol A

(17)

The simplest and most copious *oxygen*-containing molecule is *water,* H_2O or

$\overset{\displaystyle O}{\underset{H \qquad H}{}}$. There are a multitude of possibilities for incorporating this struc-

ture into organic compounds. The simplest is the combination of a hydrocarbon radical with the hydroxyl group, —OH.

Methyl alcohol (18), also called *methanol,* and *ethylalcohol (19)* also called *ethanol,* are both water-miscible liquids found in many intermediary compounds.

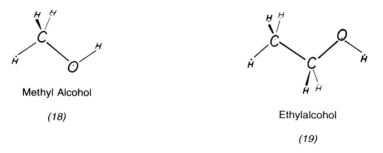

Methyl Alcohol

(18)

Ethylalcohol

(19)

There is only one synthetic polymer known that contains aliphatic bonded hydroxyl groups. *Polyvinyl alcohol (20),* which can be obtained only indirectly since the vinyl alcohol monomer does not exist, is a water-sensitive or water-soluble polymer. It has gained recognition because of its excellent barrier properties. Since it must be shielded from water, it is generally processed by coextrusion into multilayer films. Two more vinyl polymers are obtained by further chemical transformations of polyvinyl alcohol with aldehydes. Polyvinyl formal represents one component of high temperature adhesives or coatings and *polyvinyl butyral (21)* is used as an interlayer for automobile windshields to prevent injuries from flying glass debris in case of an accident.

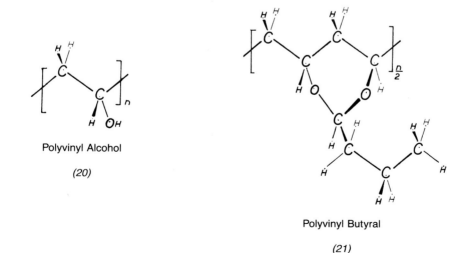

Polyvinyl Alcohol

(20)

Polyvinyl Butyral

(21)

When the phenyl radical, derived from benzene, is combined with a hydroxyl group *phenol (22)* is obtained. It is a partly water-soluble crystalline solid and is the main component of phenolics and also a raw material for many polyether, polyester and epoxy resins.

Phenol

(22)

When two hydrocarbon radicals attach to each side of the oxygen atom -ether, oxy-, or -oxide compounds are obtained. *Diethylether (23)* is a very volatile liquid once widely used as an anesthetic. The simplest aromatic ether, *diphenylether (24),* reoccurs in many high-temperature polymers.

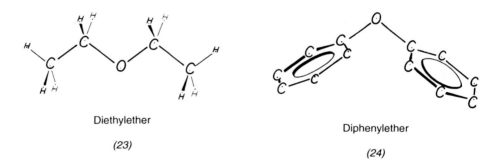

Diethylether

(23)

Diphenylether

(24)

Four representatives of polymers that contain this type of oxygen link in the polymer chain should be mentioned here: *Acetal* plastics or *polyoxymethylene (25), polyethylene oxide (26),* and *polyphenylene oxide (27),* also called *polyphenylene ether.*

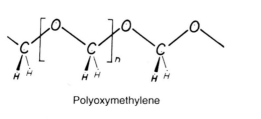

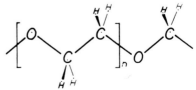

Polyoxymethylene

(25)

Polyethylene Oxide

(26)

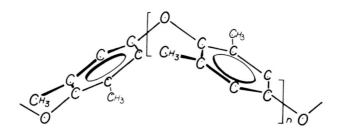

Polyphenylene Oxide

(27)

Polyoxymethylene is not perceived as a polyether. Its very high crystallinity and the formation of hydrogen bonds dominate the properties that otherwise would cause it to more closely resemble the properties of polyethylene. Polyethylene oxide represents partly water-soluble, low melting polymers. Lower molecular weight polyethers are widely used in polyurethane formulations.

The ether polymers with aromatic rings in the chains exemplify water- and hydrolysis-resistant high-temperature polymers, based on the dominance of the benzene ring.

In *methyl cellulose (28)* polymers six methyl ether groups are introduced at pending hydroxyl groups, suspending the strong hydrogen bonds that have stabilized the cellulose crystallites. The applications for these water-soluble types of polymers extend to paints and adhesives rather than to plastics.

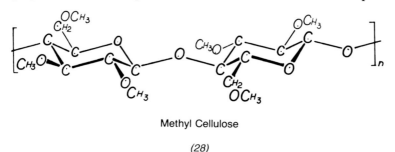

Methyl Cellulose

(28)

Two-valent *sulfur*—being similar in many respects to oxygen—will result in nearly identical structures.

Aliphatic polydisulfides (29) produce the best solvent, oil, and weather-resistant ambient temperature elastomers. *Aromatic polysulfides (30)* belong to another group of aromatic polymers, very resistant to high temperatures and chemicals. In *diphenyl sulfone (31)* the sulfur atom has a valency of six. It combines the sulfone group with two aromatic rings forming some of the strongest engineering plastics since the two neighboring phenylene rings are rigidly kept in place. The sulfone group further provides good chemical and oxidation resistance, hydrolytic stability, and reduced flammability.

Bivalent radicals may carry two hydroxyl groups resulting in glycols.

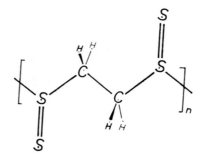

Aliphatic Polydisulfides

(29)

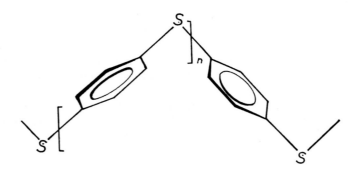

Aromatic Polysulfides

(30)

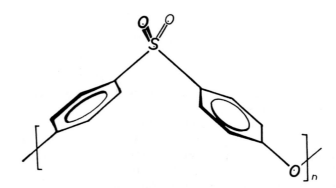

Diphenyl Sulfone

(31)

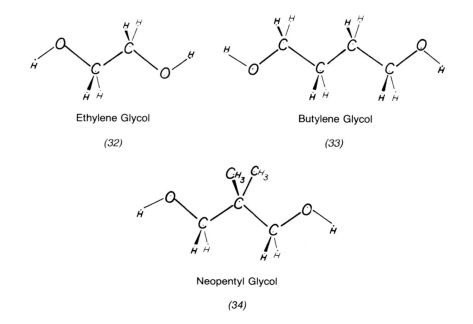

Ethylene Glycol

(32)

Butylene Glycol

(33)

Neopentyl Glycol

(34)

Ethylene glycol (32), often just called *glycol,* is a water-miscible liquid and is a reactive component of the important polyethylene terephthalate polyesters. The larger molecules, *butylene glycol (33)* and *neopentyl glycol (34),* are also important for polyesters. Many different aliphatic glycols and polyglycols, which are also polyethers, are reacted with acids, anhydrides, and isocyanates to make polyesters and polyurethanes.

As representatives from aromatic dihydroxy compounds *hydrochinone (35), biphenol (36), and bisphenol A (37)* should be mentioned since they represent important building blocks for aromatic polyesters, polyethers, and epoxies, as well as polyarylethersulfones or ketones. *Glycerol or glycerin (38)* a representative of compounds with three hydroxyl groups, is also a water-miscible, but quite viscous liquid. It is an important constituent of alkyd molding compounds and many paint and varnish resins. *Pentaerythritol (39),* with its four hydroxyl groups, represents an ideal cross-linking material. Its structure, as can be seen, resembles a cross.

Hydrochinone

(35)

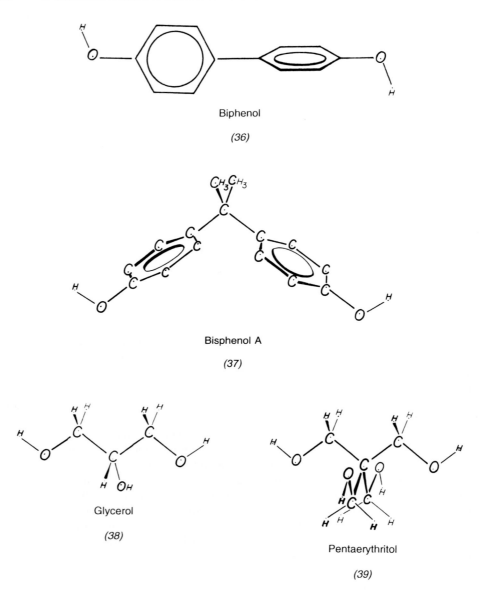

Biphenol

(36)

Bisphenol A

(37)

Glycerol

(38)

Pentaerythritol

(39)

Many molecules with more than three hydroxyl groups belong to a class of compounds generally termed carbohydrates, which are prevailingly of vegetative origin. Of the five hydroxyl groups contained in some sugars, monosaccharides, only two participate in the polymerization reaction. Two very important polymers derived mainly from glucose are *starch* and *cellulose*, which are thoroughly described in Chapter 13. Some of the remaining hydroxyl groups can be reacted chemically to modify their properties by reducing their tendency to form crystalline structures, making them melt and solution processable. The important plastics of this group are termed cellulosics.

If two hydroxyl groups are substituted for two hydrogen atoms connected to the same carbon atom an unstable compound would result. It will consolidate by splitting off one water molecule resulting in an organic compound carrying a double bonded oxygen atom. Representative compounds that contain this group—called carbonyl- or oxo-group—are designated as either *aldehydes (40)* or *ketones (41)*, depending on whether that carbon atom carries a hydrogen atom or does not.

Aldehydes   Ketones

(40)   *(41)*

Formaldehyde (42), the simplest representative, is derived from methane. With several reactive molecules it can condense to three-dimensionally, highly cross-linked resins, a process by which most of the oxygen atoms are eventually eliminated. If formaldehyde is polymerized by itself or with a co-monomer, the characteristic oxygen double bond is opened, thus forming the polyoxymethylene or acetal resins *(25)* described, on p. 62.

Only aromatic ketones, the *benzophenone (43)* derivatives, retain their ketone structure when converted to polymers. In these cases the keto-group is shielded by two phenylene rings and does not participate in any chemical reaction. Examples are the polyetheretherketone plastics.

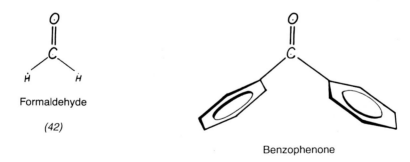

Formaldehyde

(42)

Benzophenone

(43)

Compounds having both a carbonyl and a hydroxyl group attached to the same carbon atom represent organic acids bearing the characteristic *carboxyl group (44)*. These compounds are characterized—like the inorganic acids—by their ability to easily dissociate into ions, the positively charged H^+ *cation* and the *negatively charged anion (45)*. These reactive organic acids are mostly soluble in diluted sodium hydroxide (or other bases) but only a few are readily soluble in water. In addition to their ability to form salts, other more characteristic reaction products are the polymeric esters.

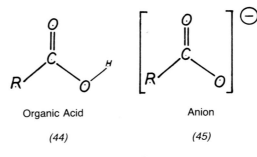

Organic Acid   Anion

(44)   (45)

Carboxyl Group

Esters form when one carboxylic group reacts with a hydroxyl group of an alcohol or glycol. A small number of free carboxylic acid groups are retained in a few polymers to impart special properties, e.g., provide solubility in slightly alkaline aqueous solutions, improve adhesion of the polymer to metal or glass surfaces, or form polymer salts leading to ionomer plastics. Although these copolymers consist mainly of ethylene chain links, the few carboxylic acid groups act as ionic, temperature-reversible cross-linking sites resulting in a plastic with exceptional low-temperature impact strength.

If the monomer contains an ester group that does not participate directly in the polymerization reaction, the ester groups remain unchanged as pendant groups. These polymers should therefore not be called polyesters. In *vinyl acetate (46)* the alcohol-derived side of the ester participates in the polymerization reaction. In *methyl acrylate (47)* it is the acid-derived side of the ester that propagates the polymer chain. This difference becomes clearer when the ester polymers are subjected to chemical hydrolysis after polymerization. In the first instance, the remaining polymer is a polyalcohol, in our example polyvinyl alcohol and in the second instance a polyacid, polyacrylic acid. Both these water-soluble polymers have special applications in sizing of textile fibers and in formulating oil- and solvent-resistant coatings.

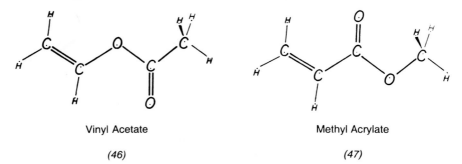

Vinyl Acetate   Methyl Acrylate

(46)   (47)

The proper designation of polyesters is furthermore used for two quite different product groups: the unsaturated thermoset polyesters and the thermoplastic polyesters. Both are obtained by linear polycondensation reactions in which the *ester group (48)* represents segments of the polymeric chain. Therefore both R_1 and R_2 must be bivalent. *p-Hydroxybenzoic acid (49)* containing

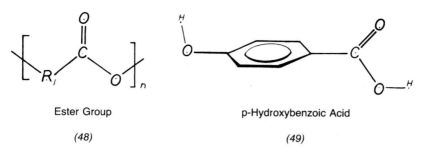

Ester Group

(48)

p-Hydroxybenzoic Acid

(49)

both a carboxyl and a hydroxyl group can polymerize by itself. Although this particular polymer is of no importance, analogous copolymers may form liquid crystal polymers.

The more prevailing polyesters derived from *dicarboxylic acid esters (50)* are formed of diacids and glycols. Some of the starting materials used for their preparation should be listed: *terephthalic acid (51)*, mainly for thermoplastic polyesters, and *phthalic acid (52)*, mainly for thermoset polyesters. Other diacids are *isophthalic acid (53)* and *chlorendic acid (54)*, which also contributes to improved chemical and flame resistance. *Maleic acid (55)* and *fumaric acid (56)* are important unsaturated diacids that contribute ethylene reactivity to unsaturated polyesters.

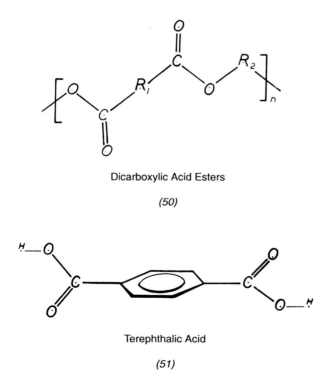

Dicarboxylic Acid Esters

(50)

Terephthalic Acid

(51)

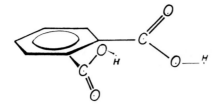

Phthalic Acid

(52)

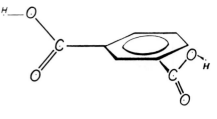

Isophthalic Acid

(53)

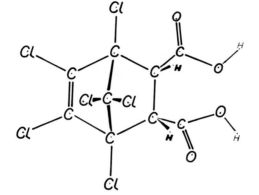

Chlorendic Acid

(54)

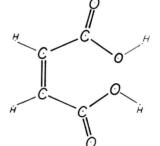

Maleic Acid

(55)

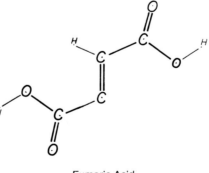

Fumaric Acid

(56)

The *polyols* that can be reacted with the above-listed diacids are *ethylene glycol (57)*, *propylene glycol (58)*, *butylene glycol* or *1,4-butane diol (59)*, and *pentaerythritol (60)*. Dicarboxylic acids have also found applications for other than polyester polymers such as the polyamides, which could contain *adipic acid (61)*. Polyimides can be obtained from *pyromellitic acid (62)* or *benzophenone tetracarboxylic acid (63)*. Polyamide-imides contain *trimellitic acid (64)*.

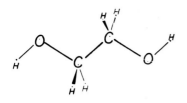

Ethylene Glycol

(57)

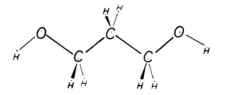

Propylene Glycol

(58)

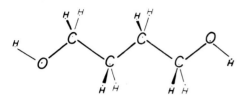

Butylene Glycol

(59)

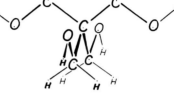

Pentaerythritol

(60)

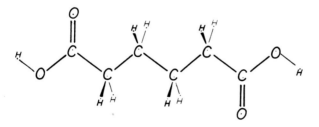

Adipic Acid

(61)

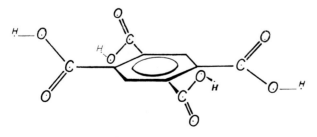

Pyromellitic Acid

(62)

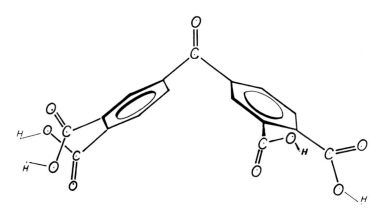

Benzophenone Tetracarboxylic Acid

(63)

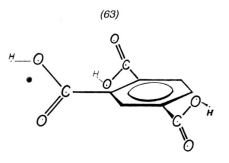

Trimellitic Acid

(64)

As raw materials, many of these diacids are employed in a more reactive form as anhydrides or acidchlorides. Two adjacently located carboxylic groups in a molecule can easily be converted to the corresponding *anhydrides (65b, 66b)*.

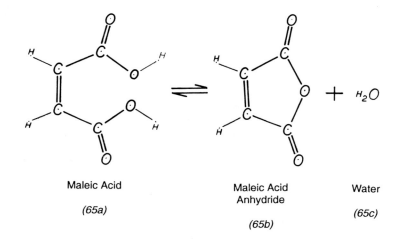

Maleic Acid

(65a)

Maleic Acid
Anhydride

(65b)

Water

(65c)

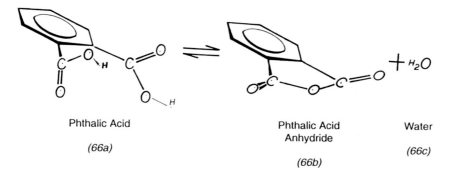

Phthalic Acid

(66a)

Phthalic Acid
Anhydride

(66b)

Water

(66c)

As is the case with water, in which one or two hydrogen atoms can be replaced by organic radicals, one, two, or all three of the hydrogen atoms can be replaced by one or several organic radicals in *ammonia* (NH_3), the simplest *nitrogen*-containing compound. To familiarize the reader with some of the possibilities, a few simple stand-ins should be listed, although most of the plastics-related representatives of this group generally derive from more complicated molecules.

The most rudimentary organic amine is *methyl amine (67)*, a gas that, like ammonia, will form a very strong base when dissolved in water. Other simple organic amines are *dimethyl amine (68)*, a secondary amine, and *trimethyl amine (69)*, a tertiary amine.

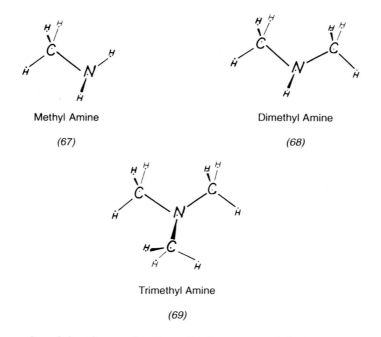

Methyl Amine

(67)

Dimethyl Amine

(68)

Trimethyl Amine

(69)

Amines employed for the production of polymers contain two or more amino groups. *Diethylene triamine (70)* is used as curing agent for epoxies and *dimethylaminoethanol (71)* for isocyanates. The most important diamino compound is *hexamethylene diamine (72)*, which, with adipic acid, forms nylon 6/6.

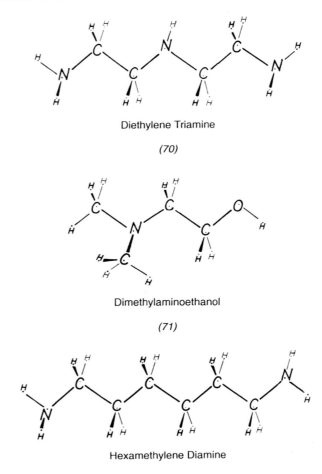

Diethylene Triamine

(70)

Dimethylaminoethanol

(71)

Hexamethylene Diamine

(72)

The aromatic compound corresponding to methylamine is *aniline (73),* a nearly water-insoluble oil. It is more widely applied as one of the difunctional *phenylene diamines (74)* (curing agents for epoxies) and the *2,4-toluene diamine (75),* a starting material for the production of diisocyanates, *methylene dianiline (76), diaminodiphenyl ether (77),* or similar compounds such as *benzidine (78)* and diamino benzidine for the preparation of high-temperature aromatic polyamides (aramids), polyimides, and polybenzimidazoles.

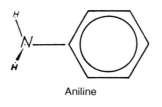

Aniline

(73)

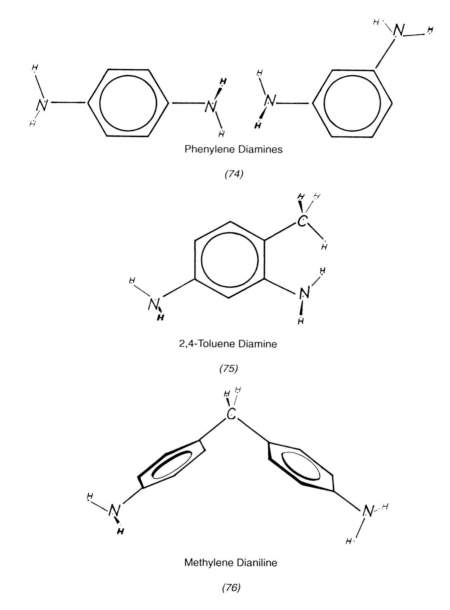

Phenylene Diamines

(74)

2,4-Toluene Diamine

(75)

Methylene Dianiline

(76)

Other important amino groups containing compounds should also be cited: *urea (79)*, the diamide of the carbonic acid, and *melamine (80)*, a cyclic trimer of cyanamide, both of which are reacted with formaldehyde to furnish thermoset amino resins.

There are two more functional groups that lend special properties to polymeric substances.

The *nitrile (81)* group is contained in many polymers including elastomers that must have good oil and solvent resistance, good weatherability, and excep-

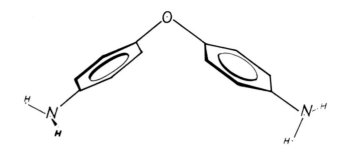

Diaminodiphenyl Ether

(77)

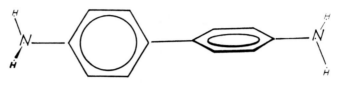

Benzidine

(78)

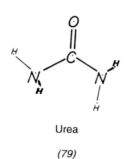

Urea

(79)

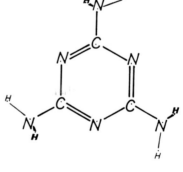

Melamine

(80)

$$R \longrightarrow C \equiv N$$

Nitrile

(81)

tional barrier properties. The main representative is *acrylonitrile (82),* which, as a homopolymer, has a rather high melting range, which poses difficulties for its melt processing. Therefore it is found alone only in fibers or in several copolymers such as the acrylonitrile–butadiene–styrene copolymers.

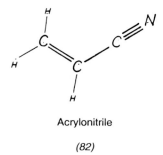

Acrylonitrile

(82)

MOLECULAR WEIGHT AND DEGREE OF POLYMERIZATION

In contrast to low-molecular-weight substances, the clear characterization of polymeric materials poses difficulties. Since they cannot be accurately identified by their melting and boiling points, as was customarily done by organic chemists, the onset of research in that field was delayed half a century.

Synthetic polymerization reactions—regardless of the type of chemical reaction employed—do not follow precisely ordered step-by-step progressions. Each individual macromolecule will have been reacted at somewhat different rates, causing the end product to be represented by a mixture of quite different high-molecular-weight polymers. The greater part of molecules may be at a certain molecular weight range, but a considerable number of them will occupy areas to either side. The ideal molecular weight distribution curve would form a bell-shaped curve as shown in Figure 4.9.

A further complication is introduced by the fact that the *average molecular weight* can be expressed in different ways, which again results in different distribution curve shapes. In the case of number averages, the smaller molecular weight fractions will be overemphasized and the high-molecular-weight fractions will dominate the track of the curve when plotted as weight averages. The reason for this is that a large number of smaller molecules will contribute an equal weight fraction as a small number of larger molecules. Only a mono-disperse polymer would possess the same number for both the weight and number averages of molecular weight.

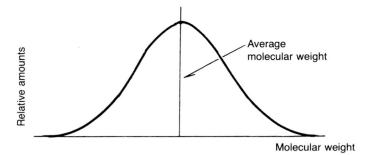

FIGURE 4.9.   Ideal molecular weight distribution curve.

If the weight average molecular weight is divided by the number average molecular weight, the *polydispersity index* is obtained. In some cases, controlled rheology resins with a narrow molecular weight distribution are preferred.

For other processes in which high-molecular-weight plastics are needed but easier processing (better flow) must be facilitated, bimodal molecular weight distribution resins are used. The lower molecular weight fraction will act as a plasticizer when extruding film, sheet, or parisons at high temperatures. The higher molecular weight fraction, on the other hand, prevents excessive sagging and ensures better mechanical properties. Bimodal polymers have gained particular importance in high-molecular-weight high-density polyethylene extrusion compounds, which now can be manufactured at lower cost in a single polymerization process. These advantages can lead to downgaging of film and pipe (bottles) products in comparison with standard plastics.

The *gel permeation chromatography* (size exclusion chromatography) method, although it cannot directly determine the absolute molecular weights, has obtained recognition as a method by which a polymer may be quickly characterized in regard to its complete molecular weight distribution, including low-molecular-weight monomers and additives. In Figure 4.10 a graph obtained by this method is shown and the individual values for the respective averages as derived by automatic data handling systems are entered.

As shown in Figure 4.11, the shape of the curves obtained from different commercial plastics may take on several forms depending on their width and the inclination of the slopes.

As pointed out before, the molecular weight of a polymer greatly influences both the mechanical properties of molded parts and the viscosity of the polymer melt. This, in turn, controls the processability of the plastic. Unfortunately both these effects are directly opposed to each other. The materials with the best properties require more effort and time to mold. Materials that easily fill mold cavities, on the other hand, may not comply with the mechanical

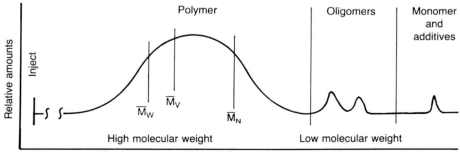

FIGURE 4.10.   Molecular weight information available with gel permeation chromatography. M_n, Number average; M_v, viscosity average; M_w, weight average. Adapted from Waters Division of Millipore Corp. Milford, MA.

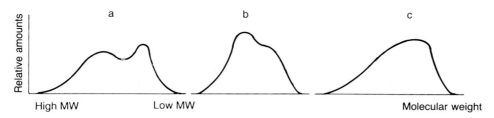

FIGURE 4.11. Molecular weight distribution curves for (a) multiple peak polymer (polymer blend, bimodal polymer), (b) shoulder at high-molecular-weight side, and (c) skewed distribution toward the low-molecular-weight side.

property requirements, especially their vulnerability to impact damages. Therefore, when a compromise has been established, it becomes essential to ensure uniformity and repeatability in regard to the originally substantiated molecular weight distribution.

Due to minor uncontrollable process parameters it is not always possible to produce batches of polymers repeatedly to exactly the same specification or to continually uphold all detailed conversion conditions over the entire time span of continuous polymerization processes. Therefore one must often blend more or less different materials to supply repeatedly a compound having a constant processing and property profile.

In some publications, the size of a polymer is described not by its molecular weight but by its *degree of polymerization*. This quantity, which reflects only the number of repeating units joined in the polymer, may illustrate the extent of polymerization better, especially with large monomer groups. Some polymers containing halogenated monomers certainly justify this kind of description since only the lineup of an adequate number of monomer units will ensure satisfactory performance of a polymer, not its high molecular weight incidentally contributed by heavy halogen atoms. As an example, each tetrabromo-bisphenol A monomer will contribute the same length as a conventional bisphenol A monomer in a polycarbonate polymer, although their molecular weights are quite different: 228 versus 544.

Usually a large number of chemically identical polymers, spread over a wide range of molecular weights, are put on the market. Probably the best example can be given by choosing polyethylene. The lowest molecular weight polyethylenes are ideal for hot melt adhesive applications whereas the highest molecular weight polyethylene (ultrahigh-molecular-weight high-density polyethylene) is required where high wear and fatigue resistance properties are demanded. The latter's processing difficulties prohibit broader application, even though these exceptional property improvements would be welcome by all users of molded products. For injection molded parts, extruded films, or blow-molded containers, the intermediate-molecular-weight plastics with selected rheological properties are generally marketed. As a general rule it may be stated that the plastic with the lowest molecular weight, as long as it provides the minimum of the end product's property requirements, represents the optimum level. The reader is also referred to Figure 14.5, p. 318.

REFERENCES

M. Chanda and S.K. Roy, Plastics Technology Handbook. Marcel Dekker, New York, 1987.

Christopher Hall, Polymer Materials, an Introduction for Technologists and Scientists, 2nd ed. John Wiley, New York, 1989.

5

Descriptions of Important Plastics

COMPARISON OF THERMOSET AND THERMOPLASTIC PLASTICS

Since the early introduction of plastics, thermosetting and thermoplastic materials have competed for use in many practical applications. In the manufacture of certain parts, the choice of which material to use has fluctuated repeatedly over the years, sometimes caused by changes in prices and sometimes by the introduction of new materials.

Although the classification of plastics into one of these two groups is not always possible, the *thermoplastics* are more easily definable. A thermoplastic material represents a linear polymeric substance that may already be compounded with several additives, usually in small amounts. This material, invariably solid at ambient temperatures, can be shaped into useful products simply by temporarily converting it into a liquid state. In most cases this is achieved by heating, causing conversion of the plastic to a viscous melt that will assume the shape of a mold cavity. On cooling it becomes a rigid structured part. Other less important ways of forming thermoplastic parts include casting catalyzed monomers into molds, cold forming sheets through the application of high forces requiring little or no heating, or dissolution of the material in solvents. The latter method can be utilized to form thin films or parts after evaporation of the solvent.

No changes in the chemistry of thermoplastic materials take place under most of those conditions, however, properties may be altered to a certain degree, depending on the way the long chain molecules settle during solidification to the final shape. These forming processes can be—theoretically—repeated indefinitely.

In processing *thermoset* plastics on the other hand, an additional chemical change occurs in the material immediately after the prepolymer is formed into the desired shape. The starting materials brought into contact with the mold surface can either be compounded liquids or solids, but they always will be able to undergo an additional chemical reaction that converts them into a solid, structural material that is nonsoluble, nonfusible, and therefore not reprocessable. These solid materials become rigid only when they are tightly cross-linked. Sparsely cross-linked substances could form flexible and even elastomeric materials, depending on the length of the chain segments between cross-links. Cross-linked polyethylene still retains its semicrystalline order and many of its properties resemble those of a higher molecular weight thermoplastic type, but will greatly improve thermal properties and stress-crack resistance. The bunched-up random coils of interconnected chains impart a high degree of extensibility to rubber.

This chemical reaction should preferably occur at the highest temperature practical to shorten the reaction time. Therefore, the molds for thermoset production are heated and a longer mold close time is allotted for proper cure. However, cooling of the molded part is not required.

Other thermoset plastics are obtained via a chemical reaction at room temperature. This method requires longer processing times and highly reactive links along the prepolymer chain. A cross-linking reaction may also be initiated by electron beam or ultraviolet irradiation excitation.

The fundamental difference between the two classes of materials should be stated. For *thermoplastics,* all chemical reactions take place in a chemical plant using reaction vessels, either in batch processing or continuous operations. This provides a means of easily determining the end point of the reaction and removing by-products or unreacted monomers.

For the production of *thermosets,* during the first part of the chemical reaction, which is also carried out in chemical plants, relatively small, mobile molecules are rapidly reacted with other small molecules to form larger molecules, called oligomers. Because of ensuing growth and cross-linking the resultant product becomes increasingly viscous. For plastics processing of thermosets, all ingredients must be present in the correct ratios and must be well dispersed. The molding compounds are prereacted with the fillers already incorporated to form B-stage resin compounds, which still must retain their ability to become liquid or fusible under the molding conditions.

During molding the highly reactive sites require an extended time for reactions to occur since they must first interdiffuse to complete the network polymer. All by-products, with the exception of a few volatile condensation products (water, ammonia, etc.), remain in the molded part. After final cure, the whole polymer structure in any molded part must be considered to be one single giant molecule. Many covalent bonds must be broken during fragmentation of such a molded part.

Concurrently, other chemical reactions may take place during cure that could be classified as thermal aging processes. Therefore—and also to keep processing time to a minimum—the reaction is seldom carried through to completion but is terminated when the mechanical and thermal properties obtained reach optimum values. For instance, a low heat deflection temperature reading is indicative of insufficient cure but low impact properties for either undercure or overcure conditions. In some cases, especially when molding thick section parts, the required in-mold cure times may be too long to maintain a swift production cycle. Therefore parts may be removed from the mold when partly cured and then postcured in an oven at a somewhat lower temperature but for a longer period of time. Recently high-temperature molding processes, requiring modifications of molds, are being introduced to shorten cure cycles and improve part quality.

In thermosets the properties of the molded parts depend on the formulation of the starting materials (only controllable by the chemical company) and the proper cure (a condition only the molder can control). Only in some specialized molding processes such as casting and reaction injection molding will the molder also be responsible for properly proportioning the various chemical ingredients.

THERMOSET PLASTICS

Properties of molded thermoset plastics are largely determined by the chemistry and structure of the resin components, by the cross-link density, and by the type of fillers or reinforcements used.

Highly cross-linked thermoset resins at one time dominated the plastics market. These *phenol–formaldehyde (1a), urea–formaldehyde (1b), and melamine–formaldehyde (1c) resins* are prepared by chemical condensation

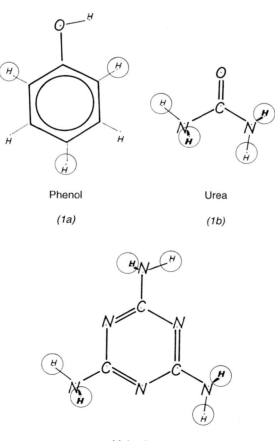

Phenol

(1a)

Urea

(1b)

Melamine

(1c)

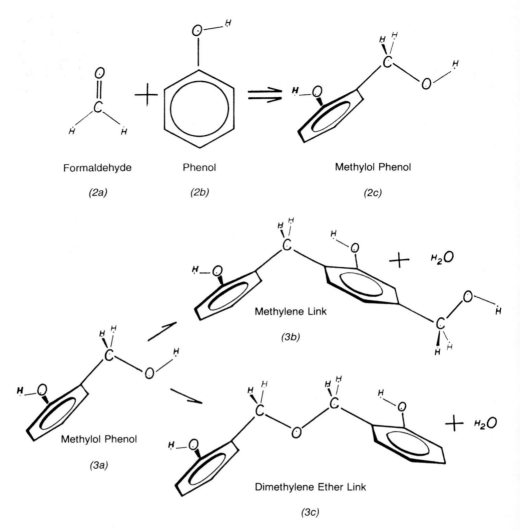

Formaldehyde

(2a)

Phenol

(2b)

Methylol Phenol

(2c)

Methylene Link

(3b)

Methylol Phenol

(3a)

Dimethylene Ether Link

(3c)

reactions of molecules—containing reactive hydrogen atoms—with formaldehyde, a very reactive cross-linking agent. The chemical formulas have the
reactive hydrogen atoms encircled. The typical reaction paths between a starting molecule and formaldehyde are shown for the example of the phenol–
formaldehyde resins *(2a, 2b, 2c)*. Two kinds of cross-linking reactions may occur
(3a, 3b and *c)*. After a series of addition and condensation reactions the highly
cross-linked phenol–formaldehyde resin is obtained under liberation of water.
In any of these three molecules more than two of the reactive hydrogen atoms
must react to obtain a three-dimensional network. However, in two-step resins
it is also possible to react first only two reactive hydrogen atoms in each
molecule and obtain a long chain, thermoplastic intermediary polymer that is
later compounded with additional cross-linking agents and finally cured like
the single-step resins. With Figure 5.1 an attempt is made to give a structural
view of such a highly cross-linked plastic.

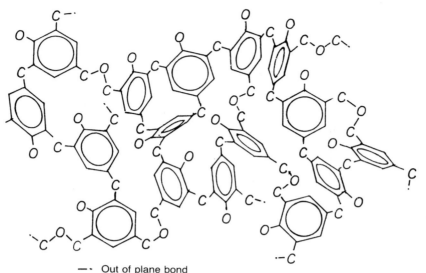

-· Out of plane bond

FIGURE 5.1. Truncated section of a tightly cross-linked phenolic plastic (hydrogen atoms omitted).

The following properties are common to these three resinous reaction products:

1. insolubility in practically all solvents,
2. will not melt when heated, and
3. extreme brittleness.

The latter fact and the very high volume shrinkage that takes place during all curing stages, especially in the case of the urea–formaldehyde products, necessitates a high percentage of filler loading. It is preferable to use a reinforcing filler for compounding into the B-stage resin, so that practically applicable structural materials can be obtained. The basic neat resins are used only in very thin layers for coatings or adhesives. A B-stage resin is still melt processable or soluble in good solvents, whereas the C-stage is referred to the fully cured resins.

The advantages of these thermoset resins are their very good compressive strength, thermal stability, dimensional stability, creep resistance, hardness, stiffness, and chemical inertness. Properties often limiting applications are low impact strength and sudden cracking when overstressed (brittle failure rather than ductile deformation).

A brief digression to *cost considerations* is inserted at this point. Although the cost of any material is not necessarily a decisive factor for pricing functional parts, the low cost of the old established thermoset molding compounds should always be taken into consideration. As a guide, a rough tabulation comparing engineering plastics with some metals they could potentially replace is given in Figure 5.2 on a price per volume basis. This relationship is

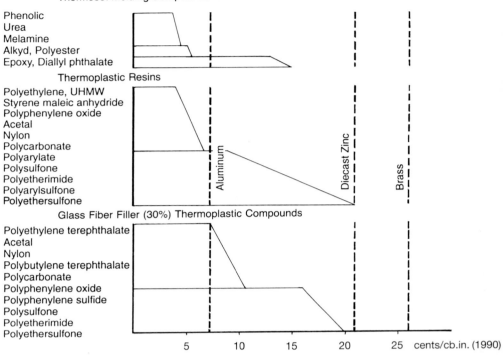

FIGURE 5.2.   Cost comparison for different engineering materials.

much more realistic since plastic parts will not reach the immense weight of comparatively sized metal parts.

At the other extreme in regard to cross-link density, a member of a thermoplastic family can be transformed into a thermoset plastic. Melt processable ordinary polyethylene can be converted to a loosely *cross-linked polyethylene* plastic either by peroxides or by irradiation. Most of the properties of these products very much resemble those of regular polyethylene with only a few important exceptions. Since cross-linked products do not melt when heated, they retain some mechanical strength up to the melting range of polyethylene, and their approximate shape right up to the point of chemical decomposition. Formed products are also much less prone to creep, wear, and environmental stress-cracking.

A very sparse cross-linking may not become sufficient to convert the thermoplastic resin into a gel structure. Under those conditions the resulting plastic would behave just like a much higher molecular weight thermoplastic. Reactions of this kind are employed in the production of extrusion and blow molding compounds.

In regard to volume usage, the cross-linkable, *unsaturated polyester resins* (4) occupy a leading position. Chemically they are composed of monomers similar to those found in some thermoplastic resins. The majority consist of a low-molecular-weight copolyester of phthalic acid and an unsaturated diacid,

generally maleic or fumaric acid, reacted with a dihydric or polyhydric alcohol.
A low viscosity resin is obtained when the polyester is dissolved in monomeric
styrene

PG　Propylene glycol
MA　Maleic acid
FA　Fumaric acid
PA　Phthalic acid
IPA　Isophthalic acid
S　Styrene

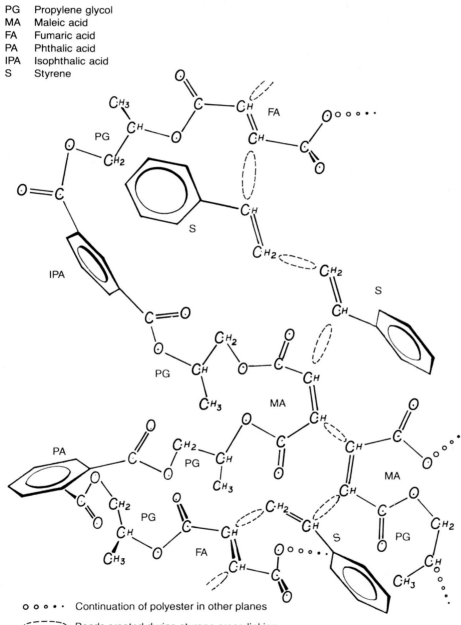

o o o • •　Continuation of polyester in other planes

　Bonds created during styrene cross-linking

Unsaturated Polyester Resin

(4)

The illustration tries to show how the ethylene linkages in a typical unsaturated polyester resin react with monomeric styrene to form a three-dimensionally interconnected network. These two components are finally chemically polymerized and cross-linked with the help of a peroxide catalyst and an amine or metal ion-containing activator. Since cured, unfilled resins are unsatisfactory because of high volume shrinkage, they are applied as composites only with reinforcing glass fibers and mineral fillers.

A great variety of compounds can be substituted for both the dicarboxylic acid, the unsaturated diacid, the glycol, and the solvent monomer, producing a wide spectrum of properties. Higher quality unsaturated polyesters involve some of the higher cost materials: isophthalic or terephthalic acid, chlorendic anhydride, bisphenol A, and vinyl toluene. Chemical formulas for many of these chemicals are shown on pp. 59 and 66–70.

Epoxy resins are superior to the above-cited unsaturated polyesters since their volume shrinkage during cure is greatly reduced. They also possess somewhat higher temperature capabilities. The reactive glycidyl radical *(5)* containing the reactive oxirane group can be incorporated by a reaction of epichlorohydrin *(6)* with many aromatic hydroxy- or amino-groups carrying compounds. Just two representatives should be noted here: diglycidyl ether of bisphenol A *(7)* and tetraglycidyl methylene dianiline *(8)*.

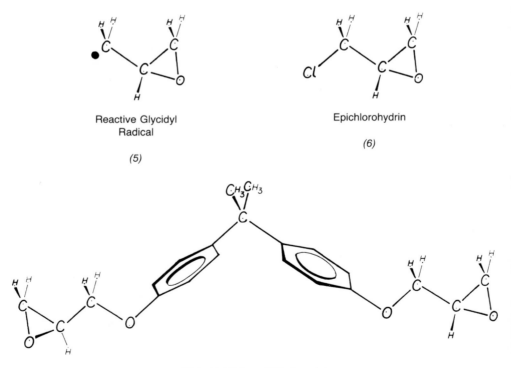

Reactive Glycidyl
Radical

(5)

Epichlorohydrin

(6)

Diglycidyl Ether of Bisphenol A

(7)

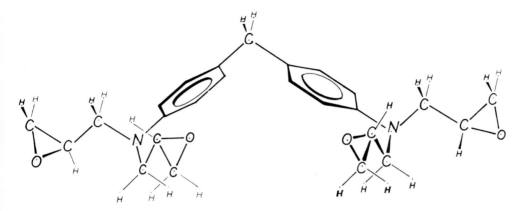

Tetraglycidyl Methylene Dianiline

(8)

The chemistry of cure of epoxies is different since reactivity is based on a ring-opening reaction (9a,b). When comparing properties of epoxides with other thermoset plastics, the extensive variability of these resins—brought about by the great variety of curing agents in use—must always be kept in mind.

Bismaleimides are processed similarly to epoxies but they cross-link in combination with other unsaturated monomers at both ends of a typical methylene dianiline-based bismaleimide monomer (10). They generally provide better high-temperature properties up to the 400–450°F (200–230°C) range.

Resins based on the combined structure of unsaturated polyesters and epoxy resins have gained importance as *vinyl ester resins (11)*. The chemical composition of some of them is based on the condensation products of bisphenol A epoxy resins with methacrylic acid. These unsaturated carbon double bonds react with styrene and peroxide catalyst systems laidup in a mold. They combine ease of processing similar to that of the glass fiber-reinforced unsaturated polyesters with the superior chemical and thermal resistance of epoxy resins. Their mechanical properties are also mainly dictated by the type, orientation, and weight fraction of the reinforcing fiber materials used.

Although their volume shrinkage during cure is less than that of the unsaturated polyester resins, the incorporation of inorganic fillers, thickeners, and dissolved thermoplastics further reduces shrinkage. Formulations having such special properties are termed—as in the case of ordinary thermosetting polyesters—low-profile resins. This is also of importance in the production of elevated temperature, pressure molded parts. During fast cure at elevated temperature, the thermoplastic resin drops out of solution occupying part of the resin matrix, thus checking shrinkage and keeping distortions to a minimum.

Diallyl phthalate resins (12) represent one of the older thermoset resins based on unsaturated phthalate esters. The chemical formula is given to indicate their similarity to the other mentioned resins. Their application, however,

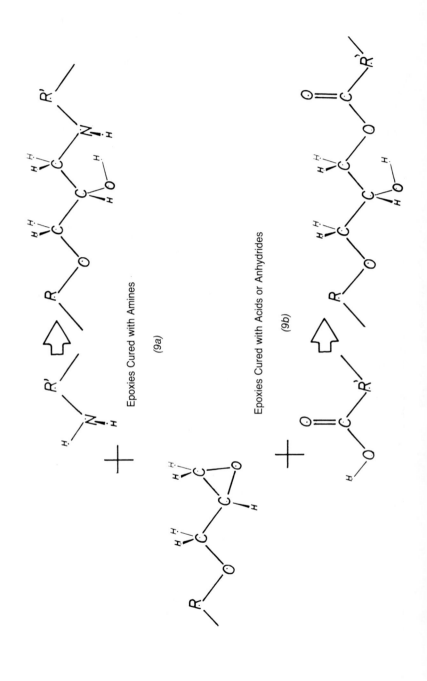

Epoxies Cured with Amines

(9a)

Epoxies Cured with Acids or Anhydrides

(9b)

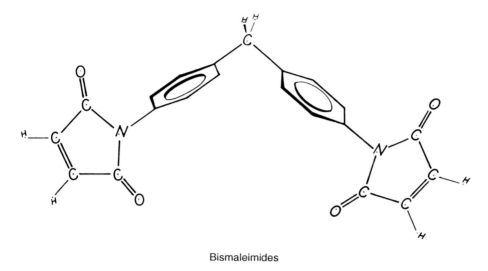

Bismaleimides

(10)

remains restricted to electrical or electronic parts, especially when the requirements surpass the capabilities of the lower cost urea– and melamine–
formaldehyde resin compounds. The molding compounds, which are highly
filled and fiber reinforced, must be compression molded.

The *polycyanate resins* or *triazine resins (13)* are represented by addition of
polymerized products of dicyanate monomers. One of them, derived also from
bisphenol A, reacts with two more cyanate groups by forming a very thermally
stable triazine ring that resembles the ring of melamine. The polycyanates are
particularly suited to replace epoxy resins for circuit board applications. Since
the resulting polymer consists mainly of densely interconnected, stable six-
membered rings, it is applied—primarily in glass fiber-reinforced laminates—
when superior electrical and thermal properties are required.

Another class of thermoset plastics comprises the *polyurethane-* and
polyurea-type resins, which are based on the high reactivity of isocyanates
with all hydroxy- and amino-groups containing raw materials *(14a–c, 15a–c).*
Difunctional isocyanates and polyols will result in thermoplastic resins but
multifunctional components must be incorporated to obtain thermoset resins.
The most common diisocyanates are the two toluene diisocyanates *(16, 17)* and
methylene diphenyl diisocyanate *(18).*

There are two classes of polyols found in most formulations, the propylene
oxide-derived hydroxyl terminated polyethers and the adipic acid-derived
hydroxyl terminated polyesters. A great number of additives are employed to
facilitate processing and to modify properties. Therefore, in many cases, a
three-component liquid mixture is dispensed into the heated molds. In spite of
higher costs, the polyureas are preferable to the polyurethane resins in automotive applications since they do not require catalysts, they react faster, and
they can withstand the conjured up temperature of 392°F (200°C) to which

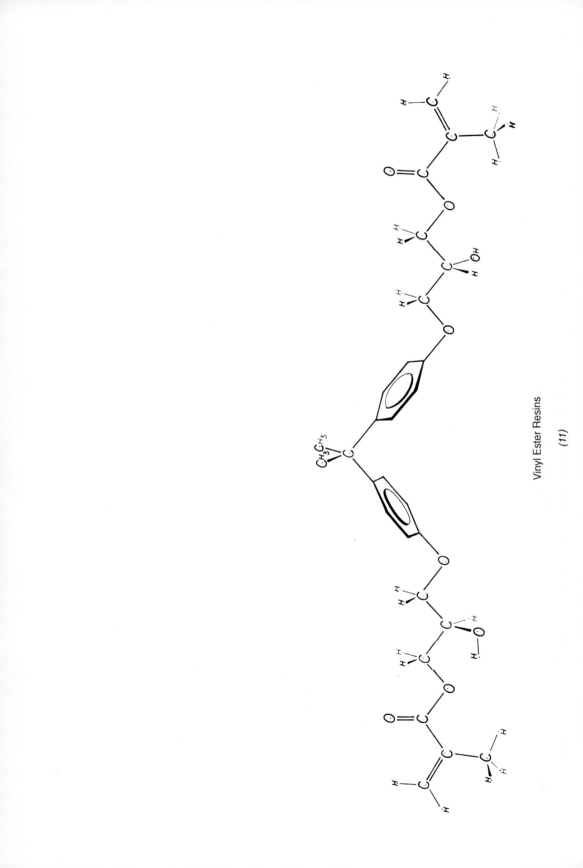

Vinyl Ester Resins

(11)

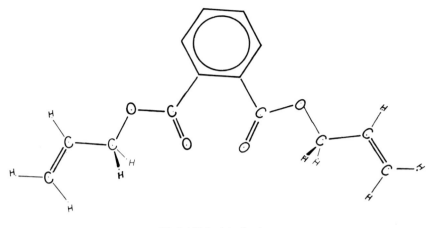

Diallyl Phthalate Resins

(12)

automotive parts may become subjected during cure of electrodeposited and spray-coated paints.

The rapid reaction of isocyanates with water according to the formula shown on p. 292 in Chapter 13 can be utilized to mold polyurethane foams, although liquids with low boiling points are preferred as blowing agents.

The primary applications of polyurethanes extend from rigid to flexible foams and to reaction injection molded (RIM) and glass fiber-reinforced structural (SRIM) parts. They also represent an important class of thermoplastic elastomers as well as true elastomers. Again because of the multitude of formulations, materials with a wide collection of specific properties are realizable. With very few exceptions, thermoset polyurethane processing is tied to the handling of multiple liquid components. Special mixing heads have been designed to quickly mix and flush out these reactive components, thus preventing resin buildup in parts of the equipment.

As with polyesters, the polyurethane bonds may also react with water under unfavorable conditions (elevated temperatures) leading to molecular weight reduction caused by hydrolysis and thus resulting in property losses.

The most widely used thermoset high-temperature matrix resin for advanced composite materials should not be left out. This mixture of low-molecular-weight reactants has been developed for fiber-reinforced prepregs. After a lengthy thermal cure process this composition will yield structural parts for aerospace applications requiring a temperature capability of over 450°F (232°C). The following monomers *(19, 20, 21)* are utilized for the formation of this cross-linked highly aromatic polyimide *(22)*.

The *silicone resins* represent a special class of thermoset plastics characterized by alternating silicon and oxygen chain members, which contribute to their high temperature stability and very broad temperature application range. Several different curing mechanisms are employed. Although the highly cross-linked silicone resins are quite brittle, cross-linked silicone elas-

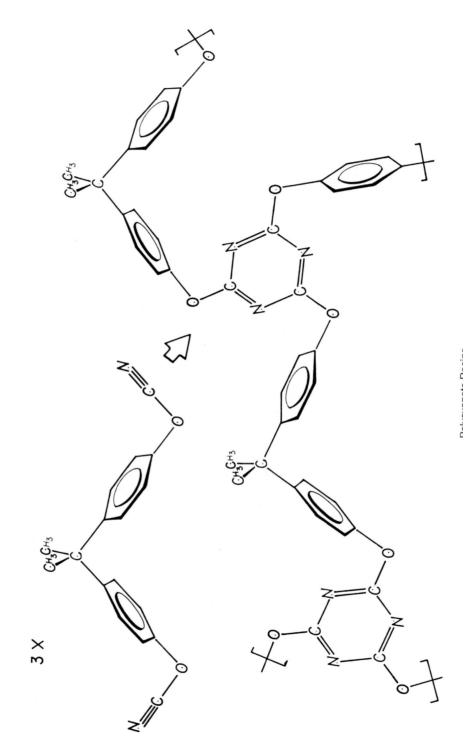

Polycyanate Resins

(13)

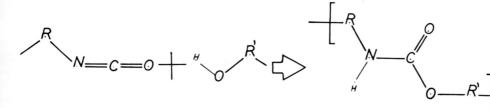

Isocyanate	Alcohol	Urethane Compound
(14a)	*(14b)*	*(14c)*

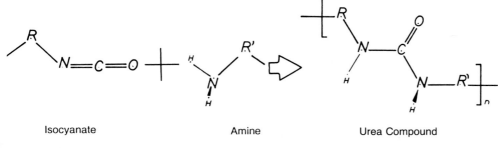

Isocyanate	Amine	Urea Compound
(15a)	*(15b)*	*(15c)*

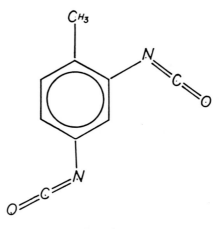

2,4-Toluene Diisocyanate

(16)

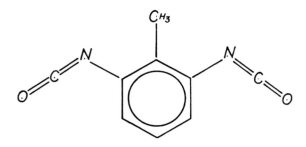

2,6-Toluene Diisocyanate

(17)

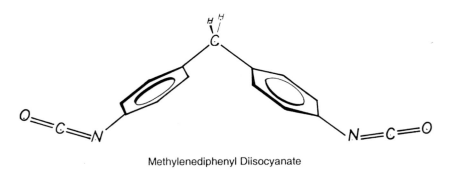

Methylenediphenyl Diisocyanate

(18)

tomers have gained wide acceptance mainly in their room temperature vul-
canizable silicone rubber form. They can be applied conveniently as one-
component viscous or thixotropic pastes that will air-moisture cure, forming
moderate strength adhesives or sealants.

Some properties peculiar to silicone rubber include their high vapor transmission rates, especially for oxygen and moisture, and their high coefficient of
thermal expansion, 4×10^{-4} in./in.·°F (7×10^{-4} m/m·K).

Interpenetrating Polymer Networks

The *interpenetrating polymer networks,* which belong to the group of thermoset
plastics, are comparable to alloys and blends of thermoplastics. These networks consist of two phases, both of which are matrices. In cases where one
phase is a plastic and the other an elastomeric material, the resulting products
can range from toughened elastomers to impact-resistant plastics. They are

recommended for applications in which noise or vibration attenuation is important. There are at least two modes for producing them: swelling of a cross-linked polymer in another monomer that is then polymerized or precipitating and cross-linking a mixture of two polymer latexes.

THERMOPLASTICS

The properties of the various *thermoplastics* are listed according to the presence of chemical elements aside from carbon and hydrogen atoms. The simple carbon chain members are grouped according to the order followed in Chapter 4, starting with simple diphatic compounds followed by aromatic and heterocylic compounds. Since the inclusion of the benzene ring structure into the chain of any polymer has a far-reaching influence on properties, these materials will be separately listed.

Aliphatic Polymers

Although *polyethylene* is predominantly used in applications in which mechanical strength is of subordinate importance, high-density polyethylene, especially in high-molecular-weight and ultrahigh-molecular-weight grades, must be considered a true engineering plastic. Its relatively low upper temperature limit of 195°F (90°C) represents the only obstacle to more widespread applications. The ultrahigh-molecular-weight grades show how increasing the molecular weight beyond the maximum tensile strength or modulus of elasticity values can still benefit performance of a thermoplastic resin. Because of the low cost of the starting material, cumbersome processing can be tolerated. A version that can be processed much more easily has recently been brought on the market. Advantage can also be taken of several unrivaled properties including excellent abrasion resistance, low coefficient of friction, and good low temperature impact strength. Applications for the high-molecular-weight grades center on pipe, sheet, and film extrusions and blow-molded containers. The good chemical and environmental stress-cracking resistance, stiffness, and toughness guarantee an extended product service life for parts utilizing these plastics.

To fine-tune the properties of polyethylene plastics—as well as all other polyolefins—three parameters must be controlled:

1. density, which is dominated by crystallinity; the crystalline content can be reduced either by favoring branching during polymerization, by the addition of comonomers, or by shortening the time span in which the formation of well-established crystallites are favored;
2. molecular weight; and
3. molecular weight distribution.

An increase in density increases stiffness, tensile strength, softening temperature, and chemical resistance, but decreases elongation, impact strength

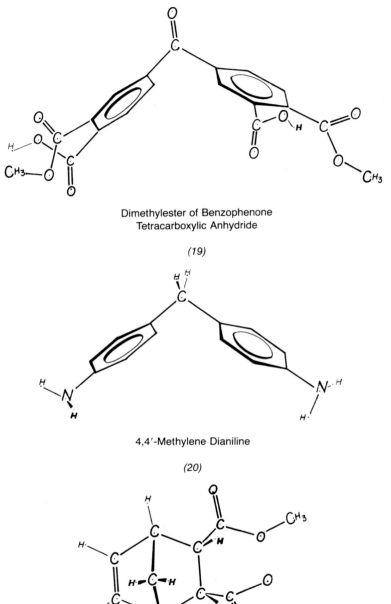

Dimethylester of Benzophenone
Tetracarboxylic Anhydride

(19)

4,4'-Methylene Dianiline

(20)

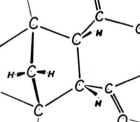

Norbornene Anhydride
Monomethyl Ester

(21)

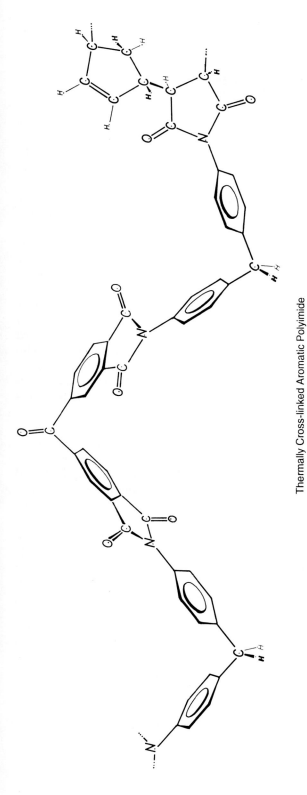

Thermally Cross-linked Aromatic Polyimide

(22)

(especially at low temperatures), and resistance to environmental stress-cracking.

An increase in molecular weight increases tensile strength, elongation, low-temperature impact strength, and resistance to environmental stress-cracking, tapering off at a certain range, but melt viscosity increases persistently.

Broadening the molecular weight distribution mainly improves flow processability by lowering melt viscosity but retaining high melt strength. It reduces impact strength but elevates resistance to environmental stress-cracking. Bimodal molecular weight distribution resins have been especially compounded for those processing conditions in which high melt strength is required (thermoforming, blow molding, and film applications).

The introduction of the lower cost process for making linear low-density polyethylene has not—and probably will not in the foreseeable future—eliminate the established high-pressure process for making low-density polyethylene. The differences in structure of both varieties result in different properties favoring certain fields of application. The monomer units in linear low-density polyethylene, which is polymerized at low temperatures, combine in a more regular fashion, resulting in straighter chain and branch backbones. This leads to the growth of more perfect crystallites, which result in greater loss of clarity and easier *uni*directional orientation. At high pressure and high temperature, the polymerization of ethylene proceeds in a more haphazard fashion, resulting in a polymer with short, random length side chains. The crystallites are therefore smaller, decreasing the scattering of light and resulting in good contact clarity. Based on its more effortless processing and *bi*directional orientation it remains the preferable material for extrusion coatings and shrink-wrap films.

The highly crystalline grades of *polypropylene* are substituted for polyethylene plastics when more rigidity and better environmental stress-crack resistance is required. The less desirable properties are their higher impact sensitivity at low temperatures, their lower thermal oxidation stability, and their lower thermal conductivity, which requires longer mold-cooling times. Advantageous are their low specific gravity of 0.90 g/cm³ and their good thermal properties, a melt temperature of 350°F (175°C) and a heat deflection temperature of 230°F at 66 psi (110°C at 455 kPa) for the unfilled plastic and up to 330°F (165°C) for the glass filled grades.

Polybutylene plastics, with their still higher melting point, exist in several crystal modifications. Since they maintain good creep properties up to 180°F (80°C), the high-molecular-weight grades are suitable for hot-water plumbing applications. Its heat deflection temperature lies also at 230°F (110°C).

The *polymethylpentene* plastic further accentuates the just mentioned differences. Although it is semicrystalline, molded parts are clear because both phases (amorphous and crystalline) have similar densities and optical properties. It has a specific gravity of only 0.835 g/cm³, the lowest of any solid plastic, and its heat distortion temperature can top 175–195°F at 66 psi (80–90°C at 455 kPa). Only its high price discourages more widespread application.

The numerous *olefin copolymers* are all characterized by their lower crystallinity, which results in a great improvement in clarity and higher flexibility. Most are also tougher. Their broad chemical composition encompasses ethylene copolymerized with the α-olefins propylene, butene, hexene, and octene, or with vinyl acetate (leading to ethylene-vinyl alcohol copolymers), methyl acrylate, and ethyl acrylate.

The *ionomers* should be dealt with in more detail since they are also applied for structural purposes. They represent copolymers of ethylene and unsaturated carboxylic acid salts. Their ionic character increases their toughness and raises their tensile strength above that of polyethylene because of the addition of stronger ionic secondary bonds acting between adjacent polymer chains. However, in comparison to the best polyethylenes, there is little gain in regard to thermal properties at the upper temperature limit or in creep properties. Water absorption is more in line with any other plastic rather than the near zero values obtained with the pure polyolefins.

Chlorine Containing Polymers

Polyvinyl chloride, one of the oldest thermoplastic resins, has claimed second place in production volume, after polyethylene, because of its many exceptional properties. Its high chlorine content (57%) ranks it first in regard to flame-retardant properties and it is often blended into other thermoplastics to improve their flame resistance. The high chlorine content also lowers their production costs. The abundance of chlorine atoms at every second link along the polymer chain contributes to secondary bonds between adjacent chains, simulating polymers of much higher molecular weight. Stretching the term crystallinity enables some researchers to designate it as having some crystalline content. Its chemical resistance is outstanding, especially against aggressive bases and acids and outdoor weathering. Mechanical strength and impact toughness are also good. Although its upper temperature capabilities are better than those of polyethylene, it is too low to be regarded as an engineering plastic. Its inferior thermal stability and the release of hydrochloric acid in case of fire are the main restraints to further extending its applications. Since it requires quite a number of additives to facilitate processing, it is usually sold to the molder already compounded.

The strong secondary bonds of polyvinyl chloride allow the incorporation of large amounts of plasticizers without losing the integrity of the plastic material. Easily processable, strong elastomeric compounds span a wide range of hardness and flexibility.

Chlorinated polyvinyl chloride is obtained by increasing the chlorine content of polyvinyl chloride. The resulting plastics have improved heat deflection temperatures under load, higher strength, and still better resistance to combustion and to chemicals. However, it is soluble in some solvents and higher in price.

Another chlorohydrocarbon polymer should be briefly mentioned here, *polyvinylidene chloride.* It is used primarily for only one purpose—a copolymer film for packaging applications. It is one of the three best plastic barrier materials.

Fluoropolymers

Polytetrafluoroethylene (23) still represents the most widely applied and lowest cost fluorocarbon-type plastic. It is used when chemical resistance, a broad temperature capability, and some special surface properties such as nonwettability and low coefficient of friction are essential. Its low strength and low creep resistance severely restrict the size of parts used in engineering applications. This limitation can be alleviated by compounding with reinforcements or by using this plastic only for thin liners or as an impregnant.

Since polytetrafluoroethylene cannot be melt-flow processed, a number of copolymers were developed that overcome the very high degree of crystallinity (90%) and allow the material to be processed similar to polyvinyl chloride. The high cost of these materials restricts them to applications in which much the

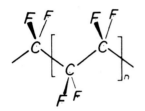

Polytetrafluoroethylene

(23)

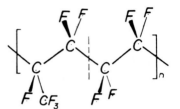

Perfluoroethylene–
Propylene Copolymer

(24)

same properties as indicated for polytetrafluoroethylene are required. Some of these are the perfluoroethylene–propylene copolymer *(24)*, the tetrafluoroethylene–perfluorovinyl ether copolymer, which has a use temperature of up to 500°F (260°C) *(25)*, and the polychlorotrifluoroethylene *(26)*.

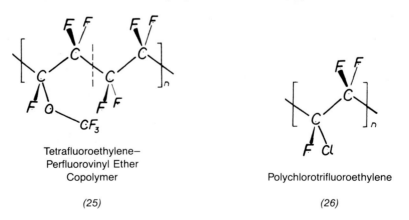

Tetrafluoroethylene–
Perfluorovinyl Ether
Copolymer

(25)

Polychlorotrifluoroethylene

(26)

Ester Polymers

Polyvinyl acetate resins are suitable only for the preparation of adhesives and coatings. Small amounts are used in the preparation of copolymers of such commodity resins as polyvinyl chloride or polyethylene. The bulk of solid polyvinyl acetate resins is chemically reacted by converting the pending ester groups into alcohol or acetal groups. The important plastic materials resulting are—as already described—polyvinyl alcohol and polyvinyl butyral.

The more extensively used polymers with pending ester groups are the *acrylics*. They comprise both the polymers of acrylic and methacrylic esters, although in most cases just the plastic consisting of the methyl ester of meth-acrylic acid is understood. The latter represents a hard, stiff plastic that distinguishes itself by its exceptionally colorless clear appearance. Other acrylics, especially the butyl acrylate monomer if copolymerized, result in soft, pliable plastics customary for many copolymers and for blends to provide elastomeric properties.

Polymethyl methacrylate's application for structural purposes is somewhat limited because of its low impact strength and susceptibility to environmental stress-cracking. In past years a number of copolymers and plastic blends have been introduced that enable the production of mechanically tough parts. Unfortunately a compromise must be made relating to optical properties.

Styrene Polymers

Polystyrene at one time represented the standard, low-cost structural thermoplastic. At a relatively low degree of polymerization in the range of 1000 monomer groups per polymer chain, it had a low melt viscosity, was easy to process, exhibited brilliant clarity, and had excellent stiffness. It is known as crystal polystyrene because of its glossy, clear appearance, although it is completely amorphous. The high modulus of elasticity (molded pieces sound like metal pieces when dropped on a hard floor) and the good heat deflection temperature of 195°F at 264 psi (90°C at 1.82 MPa) are the result of the strong secondary bonds acting between the adjacent pending benzene rings in polystyrene's structure. These bonds are inactivated under molding conditions, resulting in low melt viscosities. In spite of polystyrene's good tensile strength, its applications are curtailed because of the possibility of catastrophic failures. A sudden brittle failure could occur under impact or low strains of less than 3%. Therefore the bulk of today's polystyrene is consumed as impact polystyrene and as other styrene copolymers and polymer blends.

The addition of a polybutadiene rubber phase prior to polymerizing styrene will yield high-impact polystyrene plastics that will eliminate polystyrene's brittleness but will generally also affect its high gloss, tensile strength, rigidity, and thermal properties.

Low cost and the clarity, low melt viscosity, and good thermal stability, strength, and stiffness of polystyrene extended the search to utilize or even improve these attributes. Acrylonitrile copolymers evolved into a number of outdoor weather-resistant plastics and maleic anhydride copolymers into plastics with higher temperature resistance and higher strength. The latter component also acts as a coupling agent bonding the polymer molecules tightly to the surface of reinforcing glass fibers. This bond, as evidenced by the electron microscope photo shown in Figure 5.3, is retained even after prolonged exposure to harsh environments.

The problem with styrenics is that most practical applications demand a higher impact resistance. The possibility of accidental impact or the appearance of environmental stress crazing or cracking resulting from molded-in stresses or a flawed assembly, which could lead to disaster, led to further developments. Polymers other than polystyrene can benefit from the incorporation of an elastomeric phase, which is not limited to butadiene-carrying polymers but may include olefin and acrylate monomers. The resulting copolymers, alloys, or blends have good to excellent impact strength properties, but—unfortunately—they cost more and their other advantages are reduced.

The great variability and broad application spectrum of the very important *acrylonitrile–butadiene–styrene* plastics are based on the properties of the following three chemical components:

1. acrylonitrile, which contributes to chemical resistance, high-temperature capability, and resistance to creep deformation under load;
2. butadiene, which is instrumental in obtaining toughness because of its ability to store energy, prevent crack formation by promoting yield deformation, and curbing crack propagation; and
3. styrene, which provides a high modulus of elasticity, good processability, and low cost.

A wide-ranging diversity is achieved by changing the weight fractions of these components, which influences structure and particle size of the dispersed

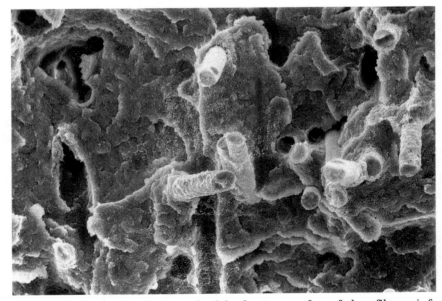

FIGURE 5.3.   Electron microphotograph of the fracture surface of glass fiber-reinforced Dylark® styrene–maleic anhydride copolymer. Courtesy of ARCO Chemical Co.

phase, or by molecular weight variations. Furthermore the compatibility and the stability of the system are of great importance. Therefore in most cases acrylonitrile is copolymerized into both phases to improve compatibility. Numerous variations can be achieved by changing the weight fractions and the sequence of incorporation of the three components. On the other hand, properties that are outside the capabilities displayed by the triangle, representing the component's properties pegged at the corners, must be acquired by other means. For example, to attain a higher heat deflection temperature, the styrene component can be replaced by α-methyl styrene. Further property adjustments can be made by alloying or blending with superior polymers.

In addition to the acrylonitrile–butadiene–styrene plastics, others, such as the olefin-modified styrene–acrylonitrile copolymers and the acrylic–styrene–acrylonitrile and rubber modified styrene–maleic anhydride plastics, are of importance. Polystyrene enthusiasts may also add the phenylene oxide-based styrenics. In most cases good resistance to creep is maintained for these derivatives too. Since the butadiene component is responsible for the lack of outdoor weather resistance, its replacement by acrylic elastomers is leading to superior weathering characteristics and improved gloss retention.

Polyoxyalkane Polymers

The *acetals* are the only thermoplastics consisting of aliphatic monomers having oxygen atom chain links. They are properly called polyoxymethylenes. Other polyoxyalkanes such as the polyoxyethylenes are soft and water-soluble, and therefore not applicable for structural purposes. The acetals' good strength, stiffness, toughness, and low creep, and the high melting points of 350°F (175°C) for the homopolymer and 330°F (165°C) for the copolymer are the result of strong hydrogen bonds and the high crystallinity of these plastics. Their modulus of elasticity is close to that of polystyrene. To achieve this high crystallinity in molded parts, mold temperature must be raised to about 185°F (85°C) to prolong the time the plastic dwells at a temperature range where the rate of crystallization is highest.

The low coefficient of friction and the resistance to water and solvents make them ideal for many smaller parts. High mold shrinkage and ensuing tendency toward warping limit their application for larger outspread parts.

The acetals' differing chemical structure reduces the suitability for an extension of their property profile by blending and copolymerizing as it is more easily done with any of the vinyl- and ester-type polymers.

Nitrogen-Containing Polymers

Of the three groups of synthetic polymers that contain nitrogen atoms as chain members in prevailing aliphatic polymeric structures, the *polyamides* have attained greatest prominence. The distinguishing characteristics of these three chemical structures are as follows (shown in *27a–c* the aliphatic chain segments are noted by zigzag lines).

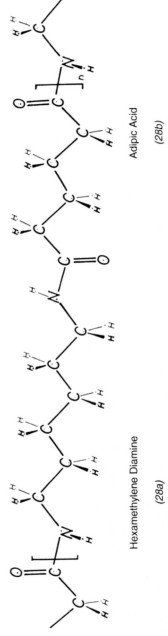

Hexamethylene Diamine

(28a)

Adipic Acid

(28b)

Nylon 6/6

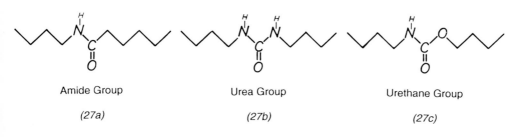

Amide Group   Urea Group   Urethane Group

(27a)   (27b)   (27c)

Polymers with the amide group are also represented in a great number of biopolymers composed of a variety of monomers, amino acids, reacted in very specific sequences. Only one such protein, collagen, will be more fully described in Chapter 13, since as a scleroprotein it represents a structurally strong, fibrous animal material. Another biogenic polyamide that has found extensive applications as binder, adhesive, and fiber and plastic is casein, which can easily be isolated out of milk whey. In all biological polyamides the interlaced aliphatic chain members between repeated amide groups consist of just one —CH_2— group but incorporate different side chains.

Some synthetic polyamides are not considered plastics although they are made up of diamines and diacids (dimerized vegetable oil acids). Their molecular weight is too low and their use is limited to curing epoxy resins or for adhesives and coatings.

Individual members of the family of nylon polyamide resins are designated by either one number or a pair of numbers separated by a vertical line. All these counts indicate the number of carbon atoms in the chain of the monomer. Those with only one number consist of only one monomer, which necessarily must have the carbonyl group $\diagdown$C$=$O on one end and the amine group $\diagup$

$\diagdown$N—H on the other. Polyamides with two numbers consist of two monomers, $\diagup$

which sequentially react with each other to form the polymer chain. The first number stands for the diamine and the second for the diacid. Therefore, nylon 6/6, the most used polyamide, is composed of hexamethylene diamine (28a) and adipic acid (28b). Its chemical name is poly(hexamethylene diamine adipate). Undulation (also seen in Fig. 6.19, p. 164) is exaggerated to emphasize that the methylene links in polyamides are not positioned along a straight line. Nylon 6, the second most important polyamide, is formed of 5-aminocaproic acid and its name is polycaprolactam (29).

Since these polyamides can be regarded as polyethylenes with interspersed amide groups, it should not come as a surprise that the properties of the higher number nylons incline more toward the properties of polyethylene. Another rule that applies here also is that chain segments with an even number of —CH_2— groups have somewhat higher melting points than the following odd-numbered polymer. The effect of the length of the alkyl chains is also ex-

TABLE 5.1. Melting Points, Heat Deflection Temperatures, and Water Absorption of Nylon Plastics

Polymer Resin	Number of —CH$_2$— Groups	Weight Fraction of —CH$_2$— Groups in Chain	Melting Temperature		Deflection Temperature under				Water Absorption (saturation at R.T.) (%)
					66 psi Load (455 kPA)		264 psi Load (1.82 MPa)		
			°F	°C	°F	°C	°F	°C	
Nylon 6	5	0.71	440	225	360	182	160	71	9.5
Nylon 11	10	0.83	375	190	302	150	131	55	1.9
Nylon 12	11	0.85	355	180	302	150	135	57	1.5
Nylon 6/6	6/4	0.71	510	265	470	243	220	104	8.5
Nylon 6/10	6/8	0.78	435	225	300	149	135	57	3.3
Nylon 6/12	6/10	0.80	410	210	—	—	—	—	2.8
Polyethylene (high density)	5000+	1	275	135	185	85	—	—	<0.01

pressed in water absorption. Table 5.1 illustrates this relationship between melting point and water absorption. Although most nylons are semicrystalline and therefore opaque, a few copolymers and elastomeric blends are amorphous and transparent.

Many applications of nylons utilize their low coefficient of friction, good abrasion, and chemical resistance. Caution should be exercised since the low number nylons in particular have a high moisture absorption that may lead to lowered mechanical properties and greater dimensional changes. One should be aware that at temperatures above about 170°F (75°C) both the mechanical properties and chemical resistance may exhibit excessive transpositions.

Because of strong hydrogen bonds, good mechanical properties are already obtained at rather low molecular weights (10,000 to 35,000), resulting in quite low melt viscosities at melt processing temperatures.

A number of thermoplastic resins have a structure similar to the nylons but carry either two nitrogen atoms or one nitrogen and one oxygen atom on either side of the central carbonyl group. Of these *polyurea* and *polyurethane* plastics only a very few grades are, like the nylons, true thermoplastics. Because of the presence of a great variability of additional functional groups, most fall more into the classification of thermoplastic elastomers (see p. 142) or thermoset plastics (see p. 91).

Ring-Containing Polymers

Since the properties of plastics materials are considerably enhanced when some of the single bonded chain members are replaced by *cyclic chain members,* plastics within this group should be dealt with separately. As elsewhere, it is difficult to try to establish a clear demarcation line. Obviously some plastics belong to both groups. A nylon resin that is half aromatic and half aliphatic, nylon MXD/6, is now actively promoted. This example confirms the

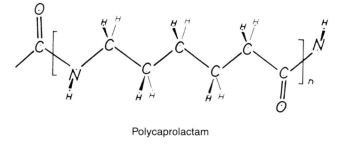

Polycaprolactam

(29)

desirability to sometimes select plastic compounds with properties that fall between the properties of aliphatic polymers—which are limited to lower temperature applications—and the properties of ring structures containing polymers—which provide a much higher temperature working range but also higher cost and more difficult processing. This goal can—as will be shown—also be achieved by blending.

In most cases the ring structures are derived from the aromatic benzene ring. Since these rings will be connected at two points to the polymer chain, designations generally found in their names refer to aryl-, phenylene-, or phthalate-, in conjunction with some other descriptive words. Similar plastics containing the six-membered saturated ring will appear with cyclohexylene or cyclohexane in their names.

Phenylene Oxide Polymers

There is also one important plastic in which the aromatic ring chain members are connected only by oxygen atoms. As is so frequently the case, several different names can describe the same compound. The polymeric main constituent in the modified *polyphenylene oxide* plastics can also be expressed as polyphenylene ether or poly(oxyphenylene). To better reflect the monomer's chemistry its name is sometimes extended to poly(2,6-dimethyl-1,4-phenylene oxide) *(30)*.

This polymer, in the form of (impact) polystyrene alloys, represents one of the engineering plastics that has attained the highest volume usage. Depending on the mixing ratios, their heat deflection temperatures are adjusted between 180 and 310°F (80 and 155°C). Their low coefficient of thermal expansion and low mold shrinkage, combined with conventional processing conditions, make them ideal for large injection molded and thermoformed parts. They have good creep and moisture resistance and adequate impact and flexural strength, but are subject to environmental stress-cracking when in contact with many solvents or oils.

When a semicrystalline resin such as nylon replaces the styrenic components in these alloys, a higher temperature and improved chemical or solvent resistance is obtained.

Aromatic Polyesters

Among the aromatic thermoplastic polyesters, those that contain aromatic dicarboxylic acids should be mentioned first.

Polyethylene terephthalate (31) when used as a structural plastic for injection molded parts usually contains glass reinforcements and nucleating agents to

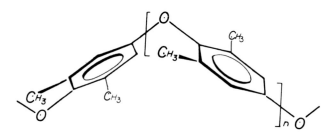

Polyphenylene Oxide

(30)

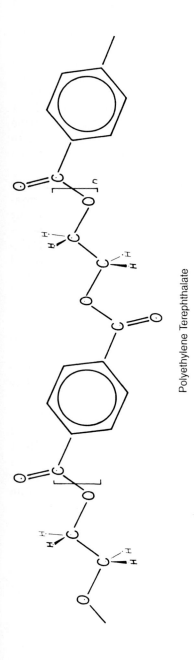

Polyethylene Terephthalate

(31)

enhance crystallization to obtain satisfactory properties. Its lower cost in comparison to polybutylene terephthalate has established a certain market for this polyester plastic. The bulk of it is still being used for fibers, films, and blown bottles. More details will be given in Chapter 6.

Polybutylene terephthalate, also a semicrystalline resin, is mainly used for injection molded parts in its mineral and glass fiber-filled forms. Its application features are good strength, stiffness, and creep resistance; its moisture and chemical resistance are fair. Limiting factors are the reduction in impact strength with increasing glass reinforcement and the danger of warpage of large parts because of its high crystalline content and nonisotropic glass fiber orientation. The material is, like all polyesters and polyimides, alkali sensitive and traces of moisture (0.005%) must be removed prior to melt processing to prevent a hydrolytic chain scission. The chemical reaction involved is shown in *32.*

Copolyesters can be obtained by either employing more than one dibasic acid, more than one glycol, or even both. Under these circumstances the formation of regularly ordered crystalline regions will be inhibited. One particular (co)polyester has gained importance when clear and colorless parts are required. This variety is polycyclohexane dimethylene terephthalate, also designated as *polycyclohexylene terephthalate (33),* carrying two six-membered rings. Although the saturated ring is not quite as rigid as the aromatic ring, it contributes to the substantially higher thermal capability, toughness, and creep resistance of these resins.

Any of these polyesters are available in an increasing number of copolyesters and blends, both neat, glass fiber-reinforced, or mineral (including mica) filled.

Polycarbonate (34) is a polyester containing two benzene rings in the two hydroxyl groups bearing part combined with the two-valent carbonic acid. The high concentration of benzene rings in the chain structure distinguishes this polyester from the previously described ones by its much higher glass transition temperature of 300°F (150°C). This is a necessity for a noncrystalline engineering plastic. Its applications are centered in fields in which excellent impact properties, tight dimensional tolerances, and good clarity and strength are required. Its weaknesses are limited solvent resistance and the possibility of crazing and environmental stress-cracking as well as its easily scratchable surface. The requirements for eliminating all moisture prior to melt processing were already stated for the other polyesters.

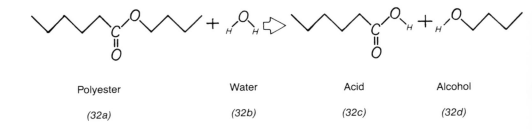

Polyester	Water	Acid	Alcohol
(32a)	(32b)	(32c)	(32d)

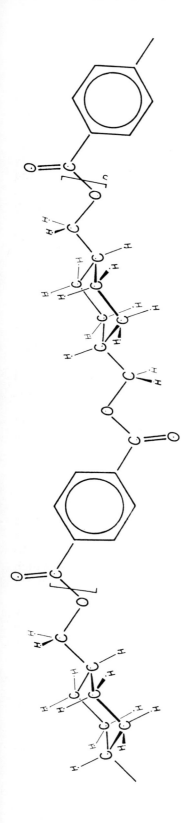

Polycyclohexylene Terephthalate

(33)

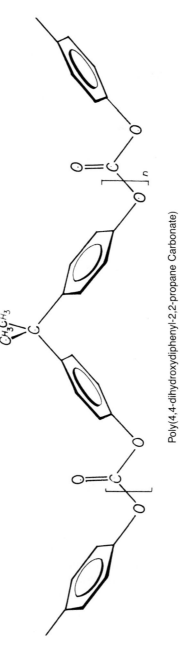

Poly(4,4-dihydroxydiphenyl-2,2-propane Carbonate)

(34)

Wholly Aromatic Polymers

The designation of *polyarylates* is generally reserved for aromatic polyesters in which both the acid and the hydroxyl group-carrying members consist of aromatic rings. However, both reactive groups may also be connected to the same phenylene ring. These chemical structures *(35–37)* illustrate some of the possible combinations and should be considered to include a variety of chemically similar components. None of these polymers is applied in large quantities and no leader has yet emerged. Some are semicrystalline but most are amorphous. Their thermal properties are so good, with heat deflection temperatures ranging from 300 to 655°F (150 to 350°C) for the liquid crystalline versions, that processing conditions border the thermal limits of conventional molding machinery.

In other wholly aromatic polymers where the aromatic rings are connected

only by —O— ether and $\diagdown C\!\!=\!\!O$ carbonyl groups, a great variety of very

high-temperature-resistant plastics can also be obtained with continuous use temperatures beyond 480°F (250°C). The polar carbonyl group imparts high temperature rigidity whereas the more flexible ether group promotes processability. A schematic structure for *polyetherketone*-type plastics should be shown and some of the names found in the literature cited: polyaryletherketone *(38)*, polyetheretherketone *(39)*, and polyetherketoneketone. Because of their semicrystalline nature, their modulus of elasticity and strength are higher than those of the otherwise similar ester and sulfone plastics.

Nitrogen-Containing Aromatic Polymers

A further increase in temperature resistance can be accomplished by adding nitrogen atoms to the chain links of highly aromatic compounds. In many instances, especially the *aromatic polyimides,* the nitrogen atom forms additional rings extending the rigid structure over two or three fused rings, which greatly improves their thermal capabilities. Their weakness lies in low ultraviolet resistance and sensitivity to alkalies. Often several different functional groups are also interposed. This variability allows a balance between property and processability. Some of the extremely rigid structured polymers must be processed like thermoset plastics since their melt temperature exceeds the decomposition temperature. The most important chain segments display the following structures as shown in *40–42.*

If these links are lined up to form polymeric chains, the following high-temperature materials, many of them already widely applied, are obtained. Some of them approach the structure of ladder polymers, which are characterized by two parallel, interconnected chain members (the rigidly connected rings that are all arranged in one plane): poly(*p*-arylamide) [poly(*p*-phenylene terephthalamide)] *(43)*, poly(*m*-arylamide) [poly(*m*-phenylene isophthalamide)] *(44)*, polyarylimides [poly(*p*-phenylene pyromelliticdiimide)] *(45)*, which is also not melt processable, and the more flexible, melt processable 5[6]amino-

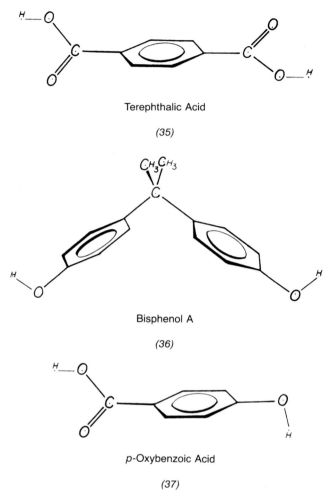

Terephthalic Acid

(35)

Bisphenol A

(36)

p-Oxybenzoic Acid

(37)

Exemplary Polyarylate Components

1(4-aminophenyl)-1,3,3-trimethylindane or 4,4´-diaminobenzophenone and benzophenonetetracarboxylic dianhydride-derived imide] *(46)*, polyarylamide-imide [poly(methylenedianiline trimelliticamideimide)] *(47)*, polyarylben-zimidazoles [poly(3,3´-4,4´-diphenyl *m*-phenylene-bisimidazole)] *(48)* and [poly(1,4-phenylene-2,6-benzobisimidazole)] *(49)]*, polyphenylquinoxaline *(50)*, polyarylbenzobisoxazole [poly(1,4-phenylene benzobisoxazole)] *(51)*, and polyarylether-imide [poly(bisphenol A 4,4´-bis-*o*-phthalic *m*-phenylene bisimide)] *(52)*.

Sulfur-Containing Aromatic Polymers

There are three classes of high-performance plastics derived from sulfur-containing aromatic compounds.

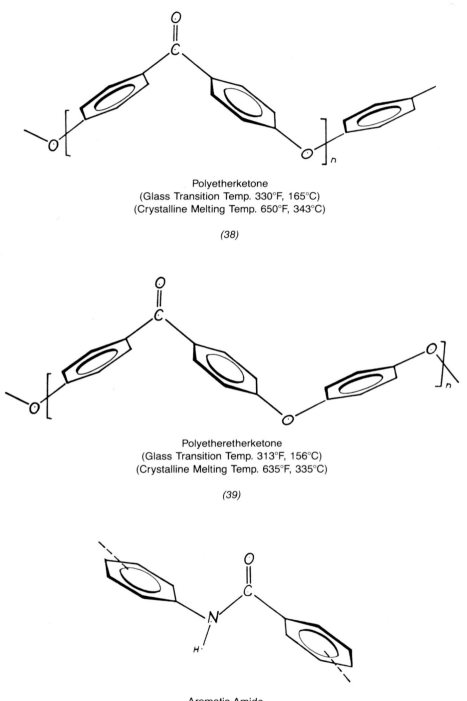

Polyetherketone
(Glass Transition Temp. 330°F, 165°C)
(Crystalline Melting Temp. 650°F, 343°C)

(38)

Polyetheretherketone
(Glass Transition Temp. 313°F, 156°C)
(Crystalline Melting Temp. 635°F, 335°C)

(39)

Aromatic Amide

(40)

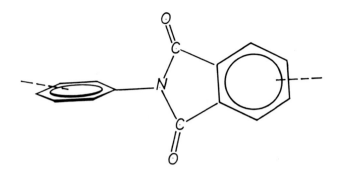

Aromatic Imide

(41)

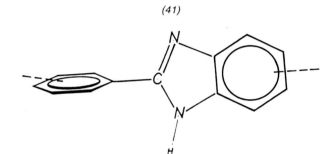

Aromatic Benzimidazole

(42)

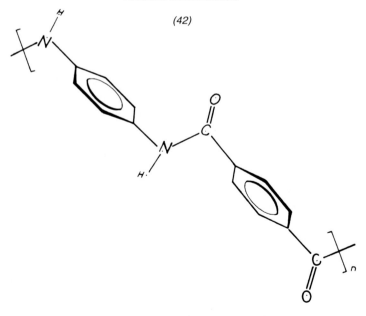

Poly(p-phenylene Terephthalamide)

(43)

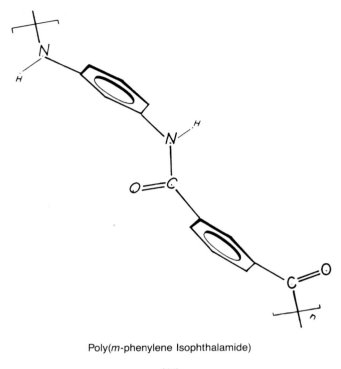

Poly(*m*-phenylene Isophthalamide)

(44)

The first are the *polyphenylene sulfides (53)*. Only their chemical structure resembles that of polyphenylene oxide; their properties are quite different. For injection molding a rather low-molecular-weight polymer is supplied. The purely amorphous resin has a glass transition temperature of only 190°F (88°C). Molded parts that can easily be converted to a material of high crystalline content have outstanding thermal and chemical stability. They are strong but must be fiber reinforced and mineral filled when impact properties are required. High mold temperatures around 300°F (150°C) will speed crystallization and ensure dimensional stability of the part as a result of the high crystalline melting point of 545°F (285°C). At quite high temperatures [above 400°F (205°C)], the molecular weight will increase because of oxidation in air, converting it into a thermoset plastic. Another high-molecular-weight version, mostly with branched chains, is noncrystalline and processable by extrusion to film and fibers. All polyphenylene sulfides are flame retardant.

Although the bond angle formed by the two rings with the sulfone group (a highly polar structure consisting of a sulfur atom with two oxygen atoms bonded to it) is 100°, the chain of the true poly(*p*-phenylene sulfone) *(54)* is too rigid, making this polymer not processable. A number of derivatives, under a great variety of chemical names, such as polysulfone, polyethersulfone, polyarylsulfone, and polyarylethersulfone, are now available that have overcome

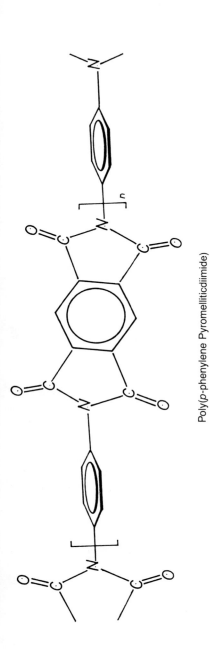

Poly(p-phenylene Pyromelliticdiimide)

(45)

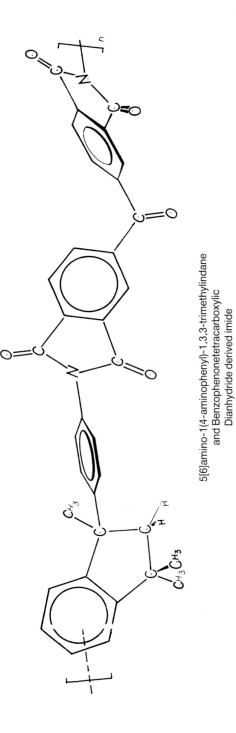

5[6]amino-1(4-aminophenyl)-1,3,3-trimethylindane
and Benzophenonetetracarboxylic
Dianhydride derived imide

(46)

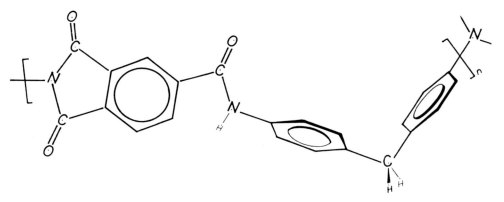

Poly(methylenedianiline Trimelliticamideimide)

(47)

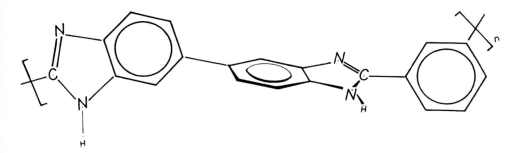

Poly(3,3'-4,4'-diphenyl *m*-phenylene-bisimidazole)

(48)

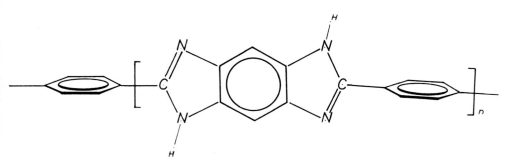

Poly(1,4-phenylene-2,6-benzobisimidazole)

(49)

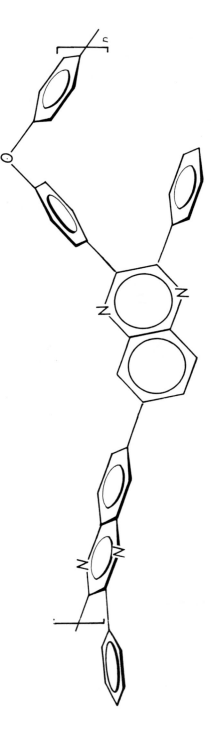

Polyphenylquinoxaline

(50)

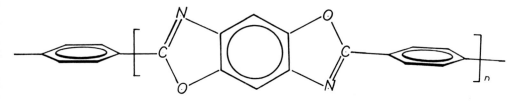

Poly(1,4-phenylene Benzobisoxazole)

(51)

this difficulty. Trade names are not listed since manufacturers keep their composition proprietary. All contain at least one ether group providing chain flexibility, which will make melt processing possible. They are all thermally very stable plastics, which, although resistant to chemical attack, will be subject to environmental stress-cracking under certain conditions. Desirable properties are their rigidity and transparency. Because they carry very few hydrogen atoms they excel in flame retardancy and low smoke generation. As in the case of the polyamide and polyamideimide plastics, the concentration of rigid structures will rank the various types of polysulfones according to their thermal capabilities *(55–58)*. The listing starts with the lowest, continuing to the highest temperature polysulfone (glass transition temperature is given below the structure). The last one *(58)* requires special equipment for processing.

In the third class of sulfur-containing polymers, the sulfur atom forms one of the heteroring atoms. *Polyarylbenzobisthiazole (59)* belongs to the most thermostable high modulus polymers. The properties of these fibers exceed in modulus and fiber tensile strength other high-temperature fibers.

Copolymers

In many cases the desired material properties cannot be obtained with homopolymers that are restricted to the described variabilities (molecular weight, selection of functional groups, and steric structure). One possibility for obtaining materials with new properties exists through copolymerizing two or more monomers with known properties. However, not all monomers can be arbitrarily reacted to form copolymers. Generally the resulting properties will be dictated by the weight fractions of the respective monomer components. Variations in properties can also be expected depending on the sequence of polymerization (e.g., whether the copolymer becomes an alternating or random copolymer or a block copolymer)

A—B—A—B—A—B—A—B—A—B—A—B—A—B—A   Alternating copolymer
A—A—A—B—A—B—B—A—B—A—A—B—B—B—A   Random copolymer
A—A—A—A—A—A—B—B—B—B—B—B—B—A—A   Block copolymer

A special arrangement in sequencing monomer units can be accomplished by graft polymerization, in which a polymer or copolymer is first formed and a

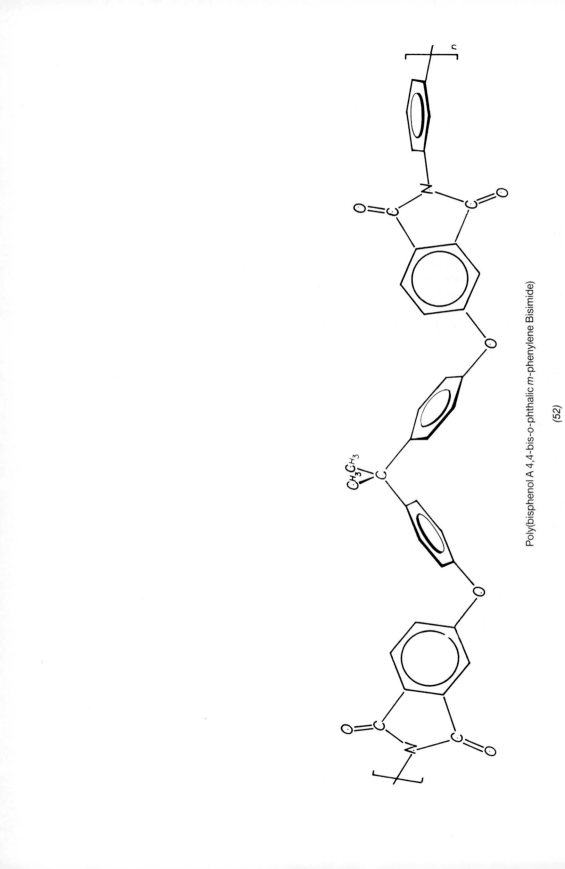

Poly(bisphenol A 4,4-bis-*o*-phthalic *m*-phenylene Bisimide)

(52)

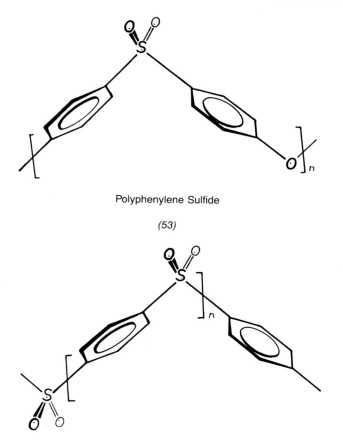

Polyphenylene Sulfide

(53)

Poly(*p*-phenylene Sulfone)

(54)

different monomer, which could form end or side chain grafts, is added afterward.

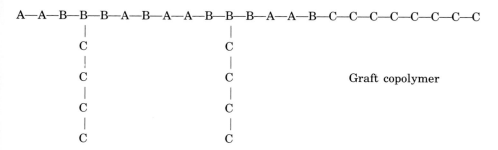

Graft copolymer

Since in this case two different kinds of polymers are combined, an inhomogeneity may develop that establishes a two-phase system similar to some inhomogeneous blends, but at a near molecular level.

Polysulfone
(Glass Transition Temp. 365°F, 185°C)

(55)

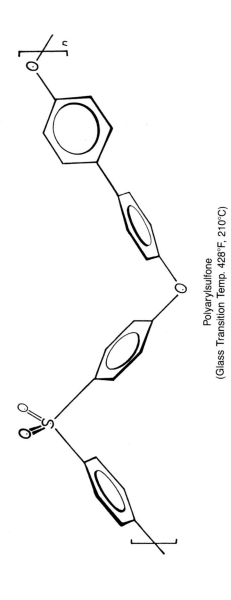

Polyarylsulfone
(Glass Transition Temp. 428°F, 210°C)

(56)

Polyethersulfone
(Glass Transition Temp. 437°F, 225°C)

(57)

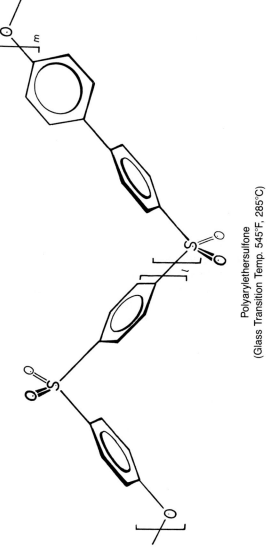
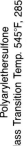

Polyarylethersulfone
(Glass Transition Temp. 545°F, 285°C)

(58)

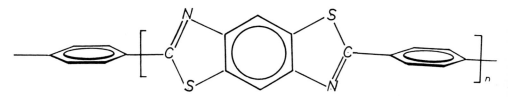

Poly(1,4-phenylene 2,6-benzobisthiazole)

(59)

The best known plastics based on this kind of copolymerization are the various acrylonitrile–butadiene–styrene plastics.

A few copolymers owe their existence to the fact that an otherwise very desirable homopolymer might not become establishable as a product because of its processing difficulties. A widespread application is ensured only after copolymerization with other easily processable polymer types. All important barrier plastics, which are applied as copolymers, fall into this category, such as polyvinylidene chloride, polyacrylonitrile, and polyvinyl alcohol.

Blends and Alloys

Alloying of metals was carried out in ancient times. In most cases several benefits could be derived concurrently. Alloys of gold were made to increase hardness; at the same time cost was reduced and corrosion resistance was maintained. Extending this technique to the quite limited number of metallic elements has resulted in a still increasing number of alloys serving present-day engineers, who must match material requirements with available material properties.

Since no basically new polymers are on the horizon, other avenues have been explored to obtain materials that more closely match the requirements for potentially high-volume applications. Based on past experience, many new polymers will not become fully commercial or will be established for only certain limited applications if they are based on compounds with too complicated chemistry. All commodity plastics are derived of very simple, low-cost intermediates.

Polymer blends, sometimes called polymer alloys, have been investigated extensively and have achieved remarkable results in many cases. Copolymerization of two or more monomers is basically restricted to related monomers that can polymerize in similar ways and under nearly identical conditions. Polymer blends, on the other hand, are prepared by mixing chemically completed polymers, mostly in the melt phase. However, most randomly picked polymer pairs would not be compatible, meaning that the combined melt would separate when cooled. This can be alleviated by the addition of a third polymer that acts as a compatibilizer. A respectable number of blends have been found that exhibit desirable properties.

However, even immiscible polymers may occasionally lead to useful plastics. Nylon and polyethylene can be formulated into a molding compound, which results in laminar structuring, leading to an appropriate application for fuel tanks, where its exceptionally low vapor transmission rate for hydrocarbons is utilized. The low rate is the result of the tortuous path these small molecules have to traverse around the nylon lamellae.

At the other extreme, a completely miscible polymer pair, polyphenylene oxide and polystyrene, represents a line of plastics that has achieved the highest market share of all polymer blends.

In cases in which polymer blends will not be at least partially miscible or could not be made compatible by the addition of block copolymers having affinity to both polymer phases, other means must be explored. A compatibilizer is being sold that can bind vinyl-type polymers with hydroxyl-group-containing polymers. This additive, required to be compounded at only 2% by weight, is a methylstyrene monomer with a pending aliphatic isocyanate group (60).

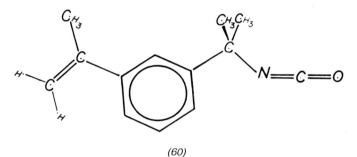

(60)

Since many blends extend over wide mixing ratios and also include semi-crystalline polymers, it becomes difficult to differentiate them all into miscible and immiscible blends. Also the number of identifiable phases may exceed two. The determination of the glass transition temperature will give the best indications. Immiscible blends will retain the glass transition temperatures of the original components, whereas miscible blends will show only one glass transition temperature at a height level reflecting the weight fractions of the components.

The first polymer blends were created out of needs similar to those described for the introduction of copolymers. One was the need to improve the impact properties of brittle thermoplastics and to prevent the sudden fracture of molded parts. An incipient crack in a brittle matrix resin can be arrested by incorporating a finely dispersed elastomeric phase. It will divert the stresses, spreading them out over a much larger volume of material. In addition to absorbing a substantial amount of energy, the distorted or broken part will probably retain enough firmness to hold an assembly of parts together. Similarly the propensity toward environmental stress-cracking can be reduced by polymer blending. In both instances similarities to the two-phase semi-crystalline plastics become apparent.

Some of the first such polymer blends, including the well-known acrylonitrile–butadiene–styrene polyblends, consist of the styrene–acrylonitrile hard phase and the acrylonitrile–butadiene elastomeric phase. Others were obtained by compounding polyvinyl chloride with chlorinated polyethylene, the latter being the elastomeric phase.

Now these types of blends have been extended to other amorphous and semicrystalline plastics including nylon, styrene–maleic anhydride, acetal, polyester, polycarbonate, polyphenylene oxide, and polysulfone. The elastomeric phase may consist of ethylene copolymers, ethylene–propylene–diene monomers, ionomers, urethanes, or soft acrylonitrile–butadiene–styrene resins.

There are still other reasons for using polymer blends. When a high-temperature resin, polyphenylene oxide, was first synthesized, its very high melt temperature prohibited processing with available machinery. Its thermal capabilities far exceeded those of other plastics used at that time, but a great demand existed for plastics with a heat resistance that would just top the crowd by 20°F (10 K). After blending polyphenylene oxide with the lower melting styrene-type polymers, their handling with conventional melt processing equipment was ensured and a significant advantage in properties was established over other styrene-based copolymers. As indicated, the two basic polymers, in this case amorphous polyphenylene oxide and amorphous polystyrene, are completely miscible and therefore form an alloy with unique properties. In those cases, where the polystyrene component is replaced by high-impact polystyrene, a blend containing an additional elastomeric phase is obtained.

In other instances a plastic material that lacks or exhibits marginal properties in some areas but adequate ones in most other areas may have to be excluded from a particular application. By changing the chemical composition of the monomer, the resulting polymer most likely would have quite different properties. However, by blending in a variant polymer resin, exhibiting an excess of the missing quality, a suitable polymer blend may be tailored. The resistance of a plastic toward creep, or its incompatibility with certain chemicals in the environment may just be such a property.

Other manifestations have been observed with polymer melts. In the case of ester-type blends, transesterification may occur with polyesters and polycarbonates at elevated temperatures. These reactions would lead to better compatibility as well. Also the rate of crystallization may be promoted by the incorporation of semicrystalline components.

Since most plastics applications are spread over a very broad basis of quite different parts, the industry has preferred to overqualify plastics materials for quality parts. The excess cost could be offset by the concomitant savings resulting from not having to make a different grade available for each of many different parts. In cases in which a single part (e.g., an automotive exterior panel) could consume a sizable share of the production volume, materials properties must be accurately tuned to requirements since material costs will become of considerably greater concern. Only blends appear to be capable of balancing price versus performance in incremental steps satisfactorily. The final selection will be made only after years of extensive use.

At present, blends of amorphous and semicrystalline materials appear to offer great promise for the fabrication of large parts. The following important properties can be controlled by combining the salient properties of each class of polymers: high impact strength and freedom of warpage contributed by the amorphous phase and stiffness, chemical resistance, and low melt viscosity provided by the semicrystalline phase.

REFERENCES

W. Brostow, Science of Materials. John Wiley, New York, 1979.

J.A. Brydson, Plastics Materials. Butterworth, London, 1975.

C.B. Bucknall. Toughened Plastics. Applied Science, London, 1977.

M. Chanda and S.K. Roy, Plastics Technology Handbook. Marcel Dekker, New York, 1987.

P.M. Hergenrother, Heat-resistant polymers. In Polymers: An Encyclopedic Sourcebook of Engineering Properties. J.I. Kroschwitz, ed., p. 427. John Wiley, New York, 1987.

M. Jaffe, High modulus polymers, In Polymers: An Encyclopedic Sourcebook of Engineering Properties. J.I. Kroschwitz, ed., p. 453. John Wiley, New York, 1987.

M.I. Kohan, Nylon Plastics. John Wiley, New York, 1973.

C.P. MacDermott, Selecting Thermoplastics for Engineering Applications. Marcel Dekker, New York, 1984.

L. Mascia, Thermoplastics: Materials Engineering, 2nd ed. Elsevier Applied Science, London, 1989.

R.B. Rigby, Engineering Thermoplastics, Properties and Applications. J. Margolis, ed. Marcel Dekker, New York, 1985.

R.B. Seymour and G.S. Kirshenbaum, eds. High Performance Polymers, Their Origin and Development. Elsevier Science Publishing, Amsterdam, 1986.

Thermoplastics and Thermosets. International Plastics Selector, San Diego, CA, 1987.

H. Ulrich, Introduction to Industrial Polymers. Hanser, Munich, 1982.

6

Structural Multiplicity of Polymeric Materials

The Amorphous State, Elastomeric Materials, and Thermoplastic Elastomers

Gaseous materials are always in a nonstructured state. They may be atomic, ionic, or molecular in nature or in some instances molecular aggregates, but the condition of their gaseous state will be unambiguous. Still rarely, under extreme conditions in certain substances, the identity of the gas or liquid states might become indistinguishable when these coexisting phases surpass a critical temperature and a critical pressure. At that point the usually visible line between the liquid and the gas will disappear. Both phases, however, will become identical even though they behave similarly to gases. Under these conditions, otherwise commonly known materials, such as water or carbon dioxide, will exhibit strange, quite extraordinary chemical and physical properties. They are now gaining interest and technical applications under the designation of supercritical fluids.

The identity of liquid and solid phases becomes blurred in many instances, particularly for polymeric materials. In all of these cases the general definition of a solid as a rigid, orderly structured material and a liquid as a mobile, randomly aggregated substance, free flowing under the force of gravity, does not hold true.

To complicate matters, crystalline materials that behave like liquids, the so-called liquid crystal materials, have also entered the realm of plastics materials. This special state of matter extends, however, only over a narrow range of temperature, close to the processing range of those materials. They are discussed in greater detail later in this chapter.

A more consequential ambivalence exists when apparently solid materials reveal the characters of amorphous materials that are expected to behave more like liquids. This behavior is generally designated as the glassy state of matter. The boundary between the solid glassy and the liquid state has been set arbitrarily at a very high viscosity of 10^{13} P (10^{12} Pa·s), with water having a viscosity of 1 cP.

The temperature at which this transition takes place is called the glass transition temperature or, not quite correctly, the second-order transition temperature. It gives some indication of whether a material will behave more like a soft, pliable material or like a rigid material, but it cannot give any information about strength. It will be fully described in Chapter 8.

Solid substances try to obtain the highest order and occupy the smallest volume possible; this is common to all materials: metals, ceramics, minerals, salts, and organic substances. There are three reasons why all materials are not in an ordered state on solidification:

1. The substances themselves are not capable of being arranged in an orderly fashion (e.g., a pile of differently sized cobblestones versus a pile of bricks).
2. Solidification (cooling) proceeds at such a high rate that there is no time for these building blocks to move to the designated lattice positions.
3. Circumstances impede the process of coordination (e.g., for the above example, not having a flat piece of ground on which to stack the bricks). A related situation for plastics would be not to have seeds from which to start the ordering of crystals or the presence of excessive amounts of impurities in the cooling melt, etc.

An additional fourth reason pertains only to long chain polymeric materials. Because of their extremely high aspect ratio (500:1) and easily flexible chain structure, it becomes impossible to accommodate 100% of these chain links in a coherent crystalline framework. All crystalline polymers should therefore correctly be designated as semicrystalline. Only a few polymeric materials that consist exclusively of straight chain members have been synthesized. These powdery materials are 100% crystalline but can be used for very few plastics applications since they will neither dissolve in solvents nor melt when heated. One example is polyparaphenylene, which can exist only as a stretched out molecule. Many times highly aromatic polyesters or polyamides are described as rod-like planar and fully extended structures, whereas in reality they still contain angular chain links facilitating deviations from a truly straight chain conformation.

Returning to the first of the three conditions just postulated, the restrictions experienced with polymeric materials can easily be recognized. Any condition affecting the regularity of a polymer molecule can suppress or prevent the formation of crystals, for example, the incorporation of chemically different chain segments in copolymers or branch points in branched homopolymers or the irregular accumulation—though chemically identical—of geometric or stereoisomer units.

The second postulate, the effect of the rate of cooling, is of preeminent importance to nearly all thermal processing steps for metals and ceramics as well as for plastics. The simplified time–temperature–transformation diagram shown in Figure 6.1 clearly describes the main phenomenon. The melting temperature and the glass transition temperature represent the boundaries between the three phases: liquid, supercooled liquid, and glass. Since the logarithm of the elapsed time is plotted on the abscissa, a high cooling rate will appear as a steeply declining line and an extended annealing process as a horizontal line. The curved line circumscribing the crystalline phase expresses the rate of crystallization very well, having a maximum located about cen-

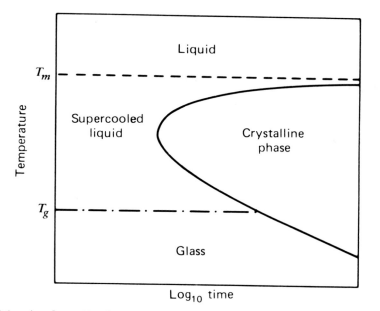

FIGURE 6.1. A schematic time–temperature–transformation diagram for solidification of a liquid; T_m and T_g denote the melting and the glass transition temperatures, respectively. From Grayson (1985), reprinted by permission of John Wiley & Sons, Inc.

trally between the melting and glass transition temperatures. For many metals these curves must be extended to accommodate a number of different crystal systems. In the case of plastics, allowances must be made for the various degrees of crystallinity as plastics cannot crystallize 100%, as well as for the differences in crystal habits.

In Chapter 4 the behavior of some hydrocarbon compounds is listed and the effect of increasing molecular weight on the melting points of these crystallizing materials is shown. Paraffin is listed as a solid material with a melting point around 140°F (60°C). It is well known that paraffin also exists as an oil, which has a low and not well-defined melting point. The molecules of both these materials may carry the same number of carbon and hydrogen atoms and may have close boiling points. The great discrepancy in regard to their tendency to crystallize when cooled indicates that significant differences in their atomic structure must exist. The readily crystallizable paraffins consist of only straight chain or normal hydrocarbons (n-alkanes) whereas liquid paraffin oil is composed of a great number of hydrocarbon isomers (same number of carbon atoms but arranged in diversely branched forms).

Both kinds will exist in the melt of mobile, randomly wound chains. Restrictions for the positioning of each chain link are imposed only by the bonds (angles and distances) connecting chain links and by the fact that it is impossible for two links (of either the same or a neighboring molecule) to occupy the same space. However, when thermal motion becomes reduced on cooling, only the chain links of the n-paraffin can be arranged into a three-dimensionally ordered crystal lattice. The mixture of the isomers, although also losing its mobility (increasing viscosity), will become an amorphous solid at a much lower temperature. Since the resistance to deformation depends on the prominence of van der Waals forces between adjacent molecules, the n-paraffin with its ordered and crowded arrangement of these bonds will be considerably harder (resisting deformation).

To describe these differences at the atomic level their relationship with the diamond (lonsdaleite) lattice should be discussed. During crystal growth each subsequent —CH_2— chain link of the n-paraffin will become locked-in once it occupies the positions in that grid that because of energetic conditions (greatest stability) are located in a zigzag line coinciding with the locations of the carbon atoms along a certain direction of the diamond (lonsdaleite) crystal. The carbon atoms are marked in black, the C—C bonds are highlighted, and the hydrogen atoms are marked in gray with thin black lines connecting them to the carbon atoms (Fig. 6.2).

In the case of the irregularly branched isoparaffins two conditions may occur after only a few carbon atoms start lining up in such a described pattern. If the succeeding chain link—added to the next lattice position—happens to be a side branch, crystallization will stop since there will be no other link avail-

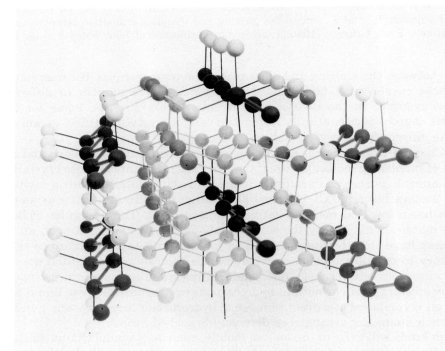

FIGURE 6.2.   Section of a paraffin crystal positioned in the lonsdaleite (diamond) crystal lattice in which all balls represent carbon atoms. Only the carbon atoms of the superimposed paraffin molecules are colored black and their C—C bonds are highlighted.

able that will allow growth to progress in that direction. If the links of the main chain succeed in continuing the growth in the correct direction, their irregular and bulky side chains will obstruct the side-by-side alignment with other zigzag chains, a prerequisite for an orderly crystal formation.

Since a larger volume must be occupied in those latter cases, this material must inevitably have a lower specific gravity. Only those links that are positioned along lines determined by crystal positions are fixed in the crystal lattice. All others should be considered potentially mobile and free to rotate within a certain arc. Many equally arranged and ordered regions must be formed to obtain the stabilizing effect of an established crystallite. In the case of branched, low-density polyethylene an appreciable number of branches will diminish but not completely inhibit partial crystal formation. This results in a lower degree of crystallinity and lower specific gravity for this polymer than for the much less branched high-density linear polyethylene.

Polypropylene, a polyethylene-type polymer with methyl groups at each second polymer chain link, also supports this principle. The polypropylene molecule with a random orientation of the methyl side groups cannot crystallize and is therefore also not usable as a structural material because of its low glass transition temperature of $-4°F$ ($-20°C$). Those polypropylene molecules in which the methyl groups are positioned in any orderly fashion will find a way to crystallize, even though not exactly in the same manner as polyethylene, since the pending methyl groups favor some other arrangements (see more details on p. 158).

Only the links at the end of or close to the end of the molecule can exercise a wide range of free gyration. More inwardly positioned links will have less mobility and contribute to the material's resistance to deformation, expressed in their melt viscosity.

Solid and structurally useful thermoplastic materials can be obtained only by increasing interchain bonding of any kind in the polymer. This can be achieved either by converting the polymer to an at least partially crystalline material, or by incorporating polar groups (chlorine, carbonyl, etc.) or larger ring structures (benzene rings) to raise the glass transition temperature. The material's strength or ability to support a load, will in one case be provided by the rigidly ordered forces within the crystallites and in the case of amorphous materials by the large number of nondirective van der Waals forces along the macromolecules. On the other hand, the rigidity of thermoset materials is based on strong chemical bonds.

For a certain, important class of materials, the *elastomers* or *rubbers,* the free rotation capability of the chain links must be maintained within the working temperature range. To prevent crystallization the chain regularity should be interposed either by two different side groups or by the insertion of other bonds not fitting into the zigzag pattern, such as double bonded chain links. When elastic rebound properties are required, the best such polymer is derived from the diene (compounds containing two conjugated double bonds) *isoprene (1).* When polymerized it can react in two ways, forming either 1,2- *(2a)* or 1,4-polymers *(2b).* The latter polymer occurs in nature in two stereo-isomers, the cis and trans configuration, brought about because it is impossible

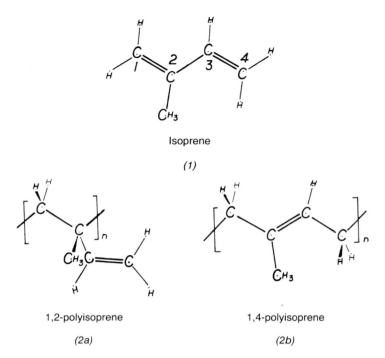

Isoprene

(1)

1,2-polyisoprene

(2a)

1,4-polyisoprene

(2b)

for the carbon atoms connected by a double bond to rotate. The trans isomer *(3)* can be contoured in a stretched out, regularly arranged form. It constitutes the hard gutta-percha material. The cis isomer *(4)* on the other hand curls up and resists proper presentation on a flat piece of paper.

These two formations *(3, 4)* are comparable to the arrangement of six atom rings containing cellulose and starch macromolecules as shown in a photo of a simplified model (Fig. 13.6, p. 296). This poly-*cis*-1,4-isoprene molecule, which can readily ball up three dimensionally because of the free rotation at all the —CH$_2$— links, is the polymer occurring in natural rubber. However, the free mobility of these groups does not become the decisive factor, as exemplified in polyethylene, which can partially crystallize although all chain links are also highly mobile. It is of greater importance that these mobile groups in *cis*-polyisoprene be prevented from arranging in a crystalline lattice because of the presence of the irregularly positioned side groups and the double bonds.

In synthetic rubber the randomness of their structure can best be obtained by random copolymerization. A very high molecular weight will promote chain entanglement, resulting in a form-stable gum stock, which still remains workable under high shear forces on heavy rubber processing equipment.

Because of the retained free gyration of the chain links, these coiled chains, when put under a tensile load, can stretch out extensively without either distorting the bond angle of 109.5° or elongating the C—C bond length. This stretching will lead to a regional orientation expanding over a short span in the chain, causing the formation of paracrystalline regions that, however, are not perfect enough to be locked in a stabilized arrangement. The achievement

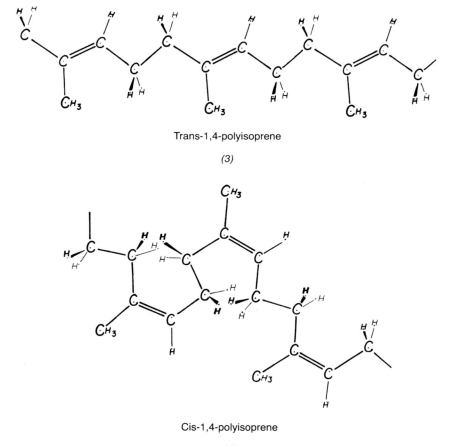

Trans-1,4-polyisoprene

(3)

Cis-1,4-polyisoprene

(4)

of a crystal-like order in natural rubber when highly stretched can be documented either by observing the increase in temperature (heat of crystallization) when a rubber band is stretched or by following the contraction of a load-supporting, extended rubber band when heated.

Since the most important property of rubber is to snap back forcefully when the load is removed, a mechanism must be incorporated for this action. Sporadic short, but strong chemical bonds must be established between adjacent chains so that they cannot slip by one another and that when the stretched out chain segments recoil, the same original shape can be reestablished. In many rubbers these bonds, called cross-links, consist of sulfur, which is reacted via a vulcanization process. However, chains can also be connected directly by reaction with peroxides and by other means.

The rubber-elastic properties of elastomers are also subject to certain environmental conditions. At low temperatures, where chain segment rotation ceases, rubbers turn into rigid amorphous materials that will shatter into pieces when subjected to impact. Only fluorocarbon and silicone class elastomers retain their elastic properties down to quite low temperature ranges. The upper temperature limit is given by the chemical decomposition range of the rubber, quite similar to corresponding thermoset plastics. The vulnerability of natural rubber to chemical deterioration at ambient temperatures, which is mainly based on the presence of unsaturated carbon–carbon bonds, can be diminished by changes in chemical composition and by additives.

A rubber-like behavior is also observed in most thermoplastic materials under conditions in which the polymer chain links pass through the state of free rotation but their viscosity is still high enough so that chain slippage remains negligible. This temperature range lies above the leathery region but below the region at which most melt processing takes place (Refer also to Figure 9.9 on p. 228). At this rubbery plateau an ordinarily rigid sheet, supported at the edges in a frame, can easily be extended. This convenient way of forming parts is utilized by thermoforming processes including vacuum forming, pressure forming, and blow molding of parisons. Excessive sagging of the heated sheet or parison would indicate that the rubbery stage had been exceeded. That forming was indeed done at a rubbery region can be shown by reheating thermoformed parts just beyond the forming temperature for a few minutes. Figure 6.3 illustrates the extent to which an injection molded preform can be extended under stretch-blow molding conditions and the extent to which this stretching can be retracted by briefly reheating the blown bottle. No change occurs in the neck region where forming was completed during the injection process.

Since processing of rubber includes the time-consuming vulcanization process, faster ways to obtain parts with elastomeric properties have been sought. By formulating a thermoplastic polymer in such a way that the rubbery region is lowered to around room temperature [with a tensile modulus of elasticity down to 1000 psi (5 MPa)], many useful *thermoplastic elastomers* have been obtained. Typically, they will consist of segmented block copolymers containing one monomer, which would result in a flexible, highly viscous liquid substance when polymerized by itself and another monomer, which would lead to a rigid thermoplastic resin. The rubbery elastomeric properties of these block copolymers are brought about by the limited compatibility between both basic polymers. The quasiliquid, soft segments of the copolymer will remain more uniformly dispersed. The hard segments of the copolymer containing the thermoplastic resin will aggregate, forming solid knots as long as the material is kept below its high glass transition temperature. These bonds will loosen up only at still higher temperatures, when the material becomes melt processable like ordinary thermoplastics. Their chemical composition is quite multifarious and extends from polyolefins, styrene butadiene, or isoprene block copolymers and polyurethanes to polyesters.

The latter, little publicized composition of a thermoplastic elastomer is detailed here as an example. The elastomer's components consist of the amorphous poly(tetramethylene oxide) terephthalate units and the crystalline poly(tetramethylene terephthalate) units as the soft and hard segments, respectively. The reader should notice that the slopes of the bonds at the ends, connecting rigid segments *(5b)*, are drawn in such a way that they can be connected to maintain the direction of the chain. The brackets are drawn

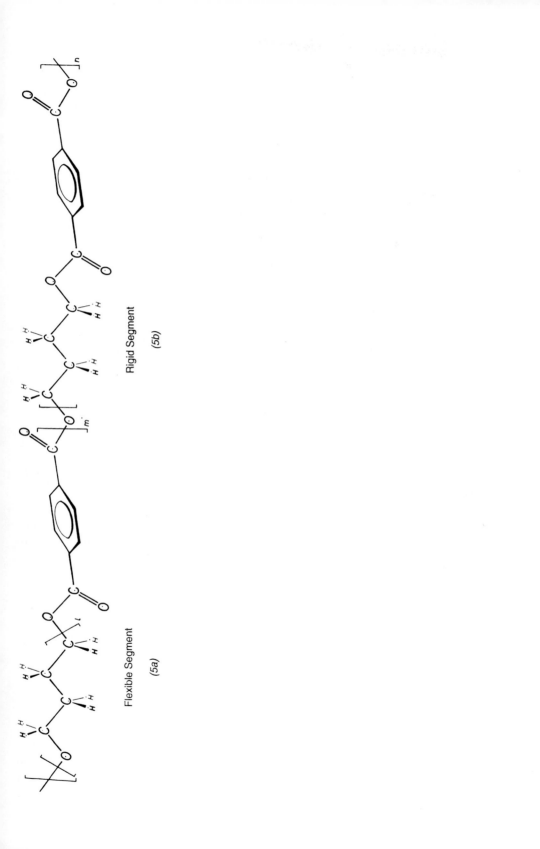

Flexible Segment

(5a)

Rigid Segment

(5b)

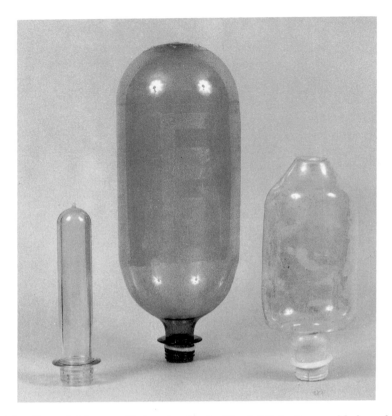

Figure 6.3.   Stretch-blow molded polyester bottle. (a) Injection molded preform, (b) blown bottle, and (c) much of the elastic stretch recovered on reheating.

parallel in the drawings. On the other hand, the flexible segments *(5a)* cannot be added in a straight fashion but must be twisted to join them at the bracketed locations. In addition, rotation could also occur at any or all the methylene links.

Thermoplastic elastomers are limited by their elastic temperature range, which is narrower than that of true rubbers. Furthermore they are not suited for high extensions or for applications in which a high rate of elastic recovery and good compression set properties are required. They are not available in low-hardness grades and will become disfigured during brief exposures to higher-than-rated temperatures.

Semicrystalline Polymeric Materials

The shortest linear chain molecule, ethane, will, as was stated in Chapter 4, solidify and crystallize at temperatures below $-278°F$ ($-172°C$) (see Table 4.1, p. 39). The aliphatic hydrocarbon with 29 carbon atoms in the chain should be regarded analogously. This paraffin, when very slowly crystallized out of a

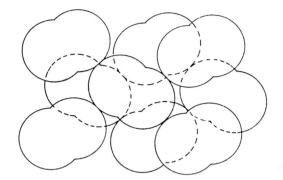

FIGURE 6.4.   Plan view of the unit cell of paraffin. The —CH$_2$— groups display correct van der Waals radii. From Evens (1964), An Introduction to Crystal Chemistry, Cambridge University Press.

diluted solution, will crystallize by laying the stretched out, still zigzag-shaped chains parallel to each other. In solution or in the melt, these molecules must be assumed to be randomly whirled, retaining the bond angle of 109.5° but varying in regard to the direction of each succeeding chain link. To accommodate more of these chains in a given volume, these zigzag planes are alternately inclined in two directions and every second chain laterally displaced in relation to the adjacent chain. Figure 6.4 shows an end-on view of the unit cell with the carbon chains all arranged parallel to the c axis. The solid-lined bisemicircles represent the space occupying hydrogen spheres around the first two carbon atoms of each paraffin molecule. The following carbon atoms up to 29 lie behind them. The spatial arrangement of the second layer of paraffin chains, hidden behind the first layer but belonging to the same crystallite, is marked with dashed lines where hidden. It is important to note that each chain is surrounded by six neighboring chains but touched by only four.

Two facts must be recognized:

1. All carbon atoms in the chain are positioned along zigzag imposed parallel lines that could be superimposed on the lonsdaleite crystal lattice. The distance between the first and fifth carbon atoms is 0.505 nm, which is equal to the diagonal distance of the diamond unit cell. In paraffin the adjacent chains must naturally be farther apart to accommodate the hydrogen atoms between them and leave an additional margin to accommodate the spacing imposed by the van der Waals radii. Those distances are 0.497 nm (a axis) and 0.745 nm (b axis), respectively. The carbon atoms of adjacent chains are positioned at right angles to the chain axis, therefore the cross-sectional area occupied by one chain corresponds to 0.185 nm^2.

2. The smallest paraffin crystals, called crystallites, will represent rhomboid platelets with a thickness of 8.16 nm, the length of two full chains, as shown in Figure 6.5. The size of these crystals is so small that only their outline is visible with an electron microscope. The molecules cannot be identified; they are only schematically indicated on the drawing.

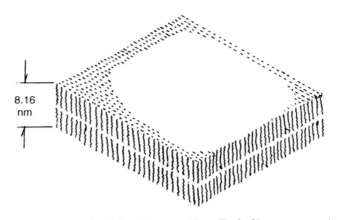

8.16 nm

FIGURE 6.5. Single crystal of $C_{29}H_{60}$ paraffin. Each line represents 29 —CH_2— groups.

Polyethylene, with a molecular weight of about 250,000, contains about 18,000 —CH_2— groups per chain. If these molecules were stretched out in analogous fashion to the paraffin molecule, they would extend over 2230 nm. It has been found that when polyethylene is crystallized slowly (24 hr) out of a 0.01% xylene solution, the resulting crystals will appear very similar to the appearance of the paraffin crystal shown in Figure 6.5. Since the thickness of the platelets is only about 10 nm—nearly the same as that of paraffin—the molecular chain must have at least been folded over 220 times. By spectroscopic means the bulk of the aligned chains can easily be identified, however, the few —CH_2— groups comprising the folds escape verification. Adjacent reentry chain folding is the logical explanation since in such a diluted solution, chain links of neighboring molecules are unlikely to be present and unlikely to be dragged in during the slow growth of the crystal. Visualizing again the structure of the lonsdaleite crystal, this type of chain fold becomes very plausible (Fig. 6.6).

Polyethylene, whether processed by injection molding, extrusion, or sintering (rotomolding) will quickly crystallize (in a small fraction of a second). This can easily be discerned by the sudden change from a clear to a white turbid appearance. Since the single crystals with the clear outlines obtained with diluted solution-grown crystals cannot be obtained with melt-grown crystals, those characteristic rhomboid or lozenge shapes are no longer clearly discernible. Only sections from the crystals will show a lamellar structure with the bulk of chain links aligned in the same perpendicular direction having a matching dimension of about 10 nm (Fig. 6.7).

Since the polymer molecules are much longer than the width of the crystallites, some kind of special arrangement of the remaining polymer segments must exist. Because of the overenthusiasm over the establishment of the adjacent reentry chain-folding mechanism for crystals grown in dilute solutions, the same structuring was first proposed for the melt crystallized polyethylene and had acquired great publicity. Naturally if that would be the case, poly-

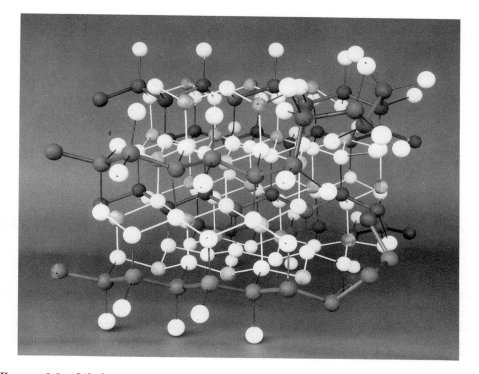

FIGURE 6.6.   Likely arrangement of chain folds in polyethylene single crystals super-imposed on the lonsdaleite (diamond) lattice.

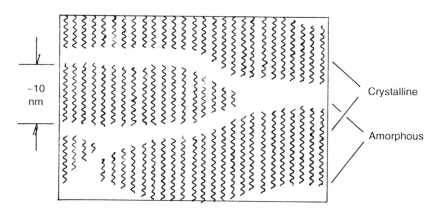

FIGURE 6.7.   Schematic sections of bulk crystallized polyethylene crystal lamellae. Chain segments are shown deliberately aligned in the paper plane. The pattern of the interspersed segments of amorphous chains cannot be identified and could be running in any direction and is therefore not shown.

FIGURE 6.8. Possible chain arrangements in amorphous regions of bulk crystallized polyethylene.

ethylene items would be no stronger than those cast of paraffin. It would be irrational to assume that every fiftieth or so chain link would be predispositioned to initiate a return bend. Where bends occur, they are most likely the result of increasing mechanical instability of an overextending straight rod. All lamellae must start out as rods and only later grow in width and breadth. Therefore other molecular arrangements, recurring at various frequencies of incidence, were proposed. Some of these are schematically indicated in Figure 6.8.

If one considers the great speed with which crystallization occurs, it is indeed very unlikely that adjacent strings of chain links will all belong to the same molecule and that in bulk materials the intertwined molecules can be instantly disentangled, isolated, and precisely knit in an orderly folded way with adjacent reentries without dragging neighboring molecules along.

It is more plausible to assume that crystallization will start in the center of what will become a lamella and work its way out to the sides until excessive entanglement with randomly distributed neighboring chain segments impedes further growth in that direction. The observation that the long-spacing will increase with higher crystallization temperature (lower viscosity and slower crystallization) and especially that, once formed, crystallites will regroup with wider average long-spacings on annealing should be taken as further evidence. It must be remembered that particularly in the case of polymeric materials, the crystallization process represents a dynamic way of action. Once under strain, formed crystal lattice ties may loosen again at a later time to reform at a more preferable site.

The beauty of the exactly arranged atoms, ions, or molecules in objects representing shapes of any one of the 32 crystal classes has fascinated humankind. However, these exactly correct crystals and the completely random arrangements of amorphous glasses represent just the two poles on a sphere, with innumerable other points on the sphere marking all the possible degrees in regularity in regard to the arrangements of atoms. This great field of randomness-in-order to order-in-randomness has just obtained recognition by physicists and is visualized in fractal concepts characterized by recursively defined geometric patterns. They are encountered in so many natural objects extending from the microscopic arrangements of thousands of ice crystals in one snow-flake to the arrangement of branches and twigs on a tree. (Just small variations in the order of overall randomness allows us to easily distinguish all the tree species.) A good perception of the structure of semicrystalline plastics can be obtained by visualizing the crystallites as leaves of a dense shrub with the amorphous fraction representing the interspersed air space.

Because the chain links in plastics are strung up, the best possible regular arrangement of the macromolecules in a crystalline form can be accomplished during processing of liquid crystalline polymers (see p. 170).

Many of these details, however, are not of any significance to the plastics practitioner since it has been established that the crystallite interfaces do not degrade the strength properties of plastics. This is not the case with metals and minerals, where grain boundaries may dominate overall properties.

A microscope equipped with polarized light optics and a heating stage can easily reveal how solidification, at a microscopic and macroscopic level, starts at nuclei points and expands from there radially in all directions. These formations consist of spherically aggregated crystallites. Not visible is the interspersed, entangled, and strained amorphous substance. These so-called spherulites (also found in some minerals and metals) will terminate their growth only when cornered by other approaching like crystal fronts. Figure 6.9 shows a schematic view of them. In the enlarged drawing the arrangement of only the crystalline sections of the chain molecules are depicted again.

In most polymers the chains are arranged tangentially to the direction of growth. Since the spherulites extend radially, the growth surface must increase rapidly. To maintain the average height of the crystallites (again about

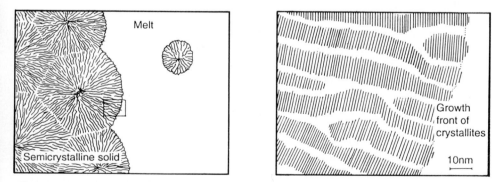

FIGURE 6.9. Spherulitic crystal growth of polyethylene (about 100× magnification). Crystallization proceeds from left to right. Lines mark centers of crystallites. Right: Enlarged schematic representation of alignment of polymer chains in lamellae or fibrils, which, however, are not restricted to one plane as shown. In certain instances these directions are spirally arranged, akin to spiral staircases.

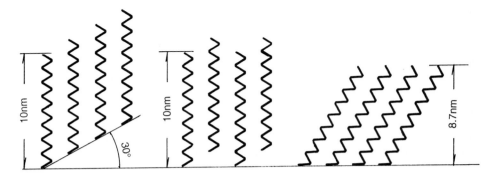

FIGURE 6.10.   Formation of different crystal habits by following the same rule. By offsetting adjacent chains by one monomer unit the obtained crystallite may assume a vertically staggered (center) or oblique structure (right).

10 nm), new lamellae or fibrils must start to fill in the voids created by the divergence of neighboring crystallites rather than by active branching.

At this point it should be noted that two and even three two-dimensional pictures will never be able to convey all details of three-dimensional structures. Similarly, the true details cannot always be established with certainty because of the flawed analytical methods applied. Some diversities should be mentioned here. Crystallites may appear to be different although built-up according to the same rule; a stepped displacement of adjacent parallel chains may differ remarkably, depending on the preference chosen when adding the next chains (Fig. 6.10). Especially in bulk crystallized materials the true crystal morphology may not be identifiable (e.g., lamellar or fibrillar arrangement).

Since the crystallite building blocks ($-CH_2-$ groups in the case of polyethylene) are connected to each other along the chain, a paracrystalline lattice formation is much more likely to occur than with atomic, ionic, or small molecule crystals. Paracrystals represent imperfectly assembled crystals in which aberrations may either be limited individually to lattice points or extended accumulatively over a longer stretch. These differences and the strain evolving between crystalline and amorphous regions constitute additional reasons why the melting points of polymeric materials are generally stretched out over a broad range in temperature. The crystalline melting points listed in the literature refer to the highest temperature, that is, when the last part of the crystallite melts.

At this point another characteristic peculiar to polymeric materials should be described. Depending on crystallization conditions, observed melting points may vary appreciably. With polyethylene the purest crystals can be obtained from a high-molecular-weight polymer crystallized under low crystallization rates at high temperatures and very high pressures. At high heating rates these crystals will show a superheating effect, which means that the polymer chain temporarily retains the order of a crystal at a temperature exceeding the melting temperature by up to 27°F (15 K). A similar observation was first made on glucose crystals. In spite of their apparent low molecular weights, sugars, like polymers, also convert into highly viscous melts.

Atomic, ionic, and low-molecular-weight substances will crystallize at a fixed temperature when cooled from the melt until completely solidified. These crystals consist of two materials: the well-ordered and materially balanced regions inside the crystals and the chemically unbalanced crystal surface. For most materials these latter areas represent the weakest sites, where the majority of mechanical failures and chemical attacks will originate. To make materials more useful for structural applications several schemes for metals have been applied: reducing grain size, improving the morphology of grains, and strengthening grain boundaries by alloying. Still the best mechanical properties of metals—besides the perfect whiskers—are obtained with noncrystalline metals, glassy metal alloys, which, however, are limited in their applications because of production problems. In the field of ceramics, glassy and microcrystalline ceramics and the purely noncrystalline glasses have gained practical applications.

Long-chain polymeric substances cannot be obtained in an entirely crystalline form. Even the most ideal chain-folded polyethylene single crystal will contain at least five —CH_2— noncrystalline links at every fold (Fig. 6.6). These amorphous regions bordering the crystallites will be more subject to chemical attack and environmental stress-cracking. The amorphous fraction in bulk crystallized polymers will, however, improve mechanical properties, mainly preventing brittle failure because of the presence of a great number of interconnecting tie links.

The high melt-index polycarbonate resin may be cited as an exception as it can turn into a brittle and fragile granular material when crystallized in the presence of solvents or diluents. Higher molecular weight polycarbonate resins, however, will be strengthened when subjected to the same conditions. In the latter case much smaller crystalline regions with plenty of tie links develop.

The contributions crystalline regions are making to plastics properties are manifold. Because of rigid positioning of the chain molecules in the crystal lattice, the modulus of elasticity and the ability of the plastic to support a load at elevated temperatures are—in all cases—significantly increased. In Table 8.1 (p. 208) the melting points of crystalline polymers and the glass transition temperatures for amorphous polymers are listed. The great difference in those temperatures can give a rough indication of how crystallinity can maintain form stability up to much higher temperatures. For other properties such as toughness no general rule has been established. Many properties are excessively affected by the size of the crystallites and the conformation in the amorphous interphase. Reducing the size of spherulites by the addition of nucleating agents without altering other crystallization conditions will result in improved mechanical properties, both in higher strength and greater ultimate elongation. This size effect is paralleled for the grain-size effect in metals.

Because of the presence of both crystallized and amorphous regions in crystallizable plastics, attention should be given to the *degree of crystallization*.

	Density (g/cm^3)	CH$_3$— Groups per 1000 Carbon Atoms	Crystallinity (%)
Crystalline polyethylene fraction	1.00	0	100
Linear high-density polyethylene, prevailingly crystalline	0.97	1	80
Branched low-density polyethylene, prevailingly amorphous	0.92	30	46
Amorphous polyethylene fraction	0.852	—	0

Typical rounded off values for polyethylene should be listed even though frequently different values are cited depending on the method used to determine crystallinity. The number of CH$_3$— end groups is about identical to the number of chain branches. Both will hinder crystal order. Even if an exact number could be established, it would not give a complete description of all circumstances. Variations in the size or the shape of the crystallites could influence the material's properties more than ±10% of the value for the degree of crystallinity. The well-established differences in the degree of crystallinity traceable to processing variables within the same materials are therefore of great importance. Higher rates of cooling will always reduce the degree of crystallinity.

The speed of crystallization is dependent on two phenomena:

1. the rate of forming crystalline nuclei and
2. the growth of the nuclei to crystallites.

The presence of nucleating agents will enhance the number (concentration) of crystallites. Since molecular mobility is required for crystal formation, the crystallization rate shows a distinct maximum at a temperature between the glass transition temperature and the melting temperature (see also Fig. 6.1). Strain can cause orientation of the crystallites and can influence morphology. The molecular weight and the molecular weight dispersity will also affect crystallinity. Higher molecular weight resins not only reduce the degree of crystallinity but also entail a more imperfect crystallite formation. However, in some instances the hindering effect of the end groups may result in greater disturbances when crystallizing lower molecular weight polymers.

This aspect must be discussed in more detail. Unlike the crystallization from very dilute solutions, crystallization in bulk should be regarded more as a parallel alignment of polymer chains somewhere in the melt; these chains extend in both directions until the entangled chains pushed to the sides by the growing crystal front exert enough resistance to halt further extension. This usually occurs after a distance of 5 to 10 nm in each direction is reached. The inner part of these crystallites will be in a reposed state, whereas the outer parts—and the closely surrounding amorphous areas—will be under strain, the result not only of tugging and pushing of chain segments during crystallization but also of the differences in the coefficients of thermal expansion and the differences in geometry and density of crystalline and amorphous regions. This, in turn, causes those outer regions to melt at a considerably lower temperature when reheated than will the relaxed central parts.

Since the degree of crystallinity depends on the available time for crystallization, the increased mobility at higher temperature will enable the continuation of growth in the height direction, thus widening the bands (up to about 50 nm). For the same reason, annealing (long time exposure to elevated temperature) of a quickly crystallized part or a part crystallized at a lower temperature will result in an increase in long spacing. The schematic sequence of sketches in Figure 6.11 shows that it is not necessary to completely melt the crystallites to accomplish this changeover. The center line of the long periods of the original crystallites cannot also be the center line for the long period after annealing, since some marginally developed lamellae will disappear adding substance to neighboring ones. Recrystallization will also take place in other substances such as metals, salts, and low-molecular-weight compounds, like delicate snowflakes that change to larger ice crystals in time.

Having generally examined the main factors affecting crystallization, circumstances that relate to specific long-chain polymers should now be considered.

Starting with the simplest polymer, *polyethylene,* it can be seen that crystallization is favored to such an extent that it becomes nearly impossible to obtain polyethylene that is not crystalline at room temperature. This is because the chain segments are highly regular and easily coordinatable. The chain is highly flexible, giving sufficient time for the chain links to be ordered into a lattice during the brief period between the melt and their immobilization resulting from increasing van der Waals forces on cooling. The physical arrangement of the polymer chains has already been described. Branched polyethylene does not obtain such a high degree of crystallinity because of the higher concentration of branch points and CH_3— end groups. This effect is even more noticeable for the many polyethylene copolymers that are crystalline to a much lesser degree, if at all. The zigzag chain conformation can also develop in other polymer crystals (polybutadiene, polyvinyl chloride, polyvinylidene chloride, polyvinyl alcohol, polyethylene terephthalate, polycarbonate, and nylon).

The fluorine analog for ethylene is another good example. *Polytetrafluoroethylene* is highly symmetric and has an extremely high degree of crystallinity. The zigzag conformation of the carbon atoms as seen in paraffin and polyethylene represents the most stretched-out conformation a carbon chain can attain and is therefore always sought if obtainable. In the case of polytetrafluoroethylene the larger size of the fluorine atoms (van der Waals radius 0.135 nm versus hydrogen with 0.12 nm) makes it necessary to twist each consecutive —CF_2— link slightly. This results in the structure shown in Figure 6.12, with a full turn obtained after about 13 units. Similar helix conformations are also found in other polymers, although at different turn ratios. Because of its high melting temperature, polytetrafluoroethylene is polymerized, going from a gaseous state directly into the crystalline polymer

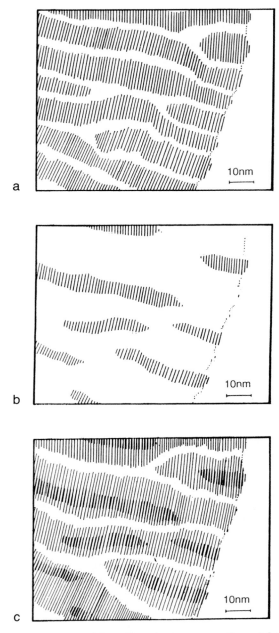

FIGURE 6.11. Recrystallization of lamellae during annealing at higher temperature. Amorphous areas are not indicated. (a) Original size of crystallites in spherulite. (b) Melting of substantial crystalline regions, not necessarily simultaneously. (c) Reformed, more stable and larger crystallites with unchanged crystalline regions highlighted.

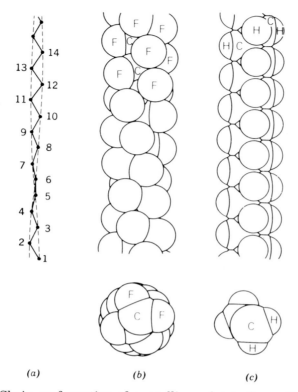

(a)   (b)   (c)

FIGURE 6.12.   Chain conformation of crystalline polytetrafluoroethylene *(a, b)* and polyethylene *(c)*. To illustrate the occupation of space all atoms are drawn according to their van der Waals radii. From C.W. Bunn and E.R. Howells. Reprinted by permission from Nature (London) **174**, 549, Copyright © 1960 Macmillan Magazine Ltd.

product. Polymers that do not have all hydrogen atoms replaced by fluorine atoms or that carry side chains are less crystalline and have lower melting temperatures, thus enabling them to be shaped by melt processing (formulas shown on p. 102).

Many ethylene-like polymers utilize monomers in which one of the four hydrogen atoms is replaced by either another atom or a group of atoms. If the first of those atoms is not hydrogen, the $CH_2{=}CH{-}$ group is called vinyl [e.g. vinyl chloride (containing one chlorine atom)]; the resulting polymer, *polyvinyl chloride (6),* is generally considered to be noncrystalline. The reason for this is the much larger size of this atom and, more significantly the nonsymmetry and the high dipole moment, which strongly connects adjacent chains preventing the slippage along chains needed to obtain a high degree of order in all three directions (crystallization). These high forces (similar to cross-links) confer outstanding coherence on this polymer, so that it still remains a high strength plastic in spite of very high plasticizer loadings. The only other polymers of similar capabilities are nitrocellulose and other cellulosics.

Polyvinyl Chloride

(6)

If one hydrogen atom is replaced by a methyl group, the properties of the resulting polymer are greatly influenced by the positioning of this methyl group in relation to the polymer chain. As in many other similar cases, the pending group (CH_3— in polypropylene) can be placed either on the right-hand side or the left-hand side in consecutively added monomer units. Certain polymerization processes yield products with random placements, also called atactic polymers, whereas some dispersed solid catalysts lead to an orderly placement of each following unit, which in turn affects the positioning of their pending groups. Polymers with random placements generally cannot crystallize and are therefore more frequently represented among the elastomeric and amorphous materials. For the polymers with orderly placements three possibilities exist:

1. All pending groups are positioned right handed, resulting in d-isomer links or
2. All are positioned left handed, resulting in l-isomer mirror imaged forms; both of these are called isotactic polymers.
3. If the pending groups become alternately positioned right and left handed, the polymer is called syndiotactic.

The chemical formulas for these polymers can be found on p. 51.

Only the *polypropylenes* with an orderly placement of the methyl group have attained importance for structural applications since only they can position chains parallel (albeit helical) in the crystallites in an orderly fashion. The chain conformations of the isotactic and the syndiotactic polypropylene molecules are shown in Figure 6.14 and Figure 6.15. Both are shown nested within the diamond lattice. By turning the diamond lattice in certain directions the various triangular, quadrangular, X-shaped, or 8-shaped chain conformations are obtained. Certain allowances naturally must be made since the bond lengths of the different atoms deviate from that involving only carbon atoms. The carbon atoms are colored black and the hydrogen atoms white. The bonds along the chain carbon atoms are highlighted. The sketches to the right indicate the projection of consecutive carbon chain atoms.

To gain a better understanding of these structures, the sequential positioning of the chain carbon atoms should be viewed from another vantage point. Three carbon atoms will always be positioned in one plane. When the fourth carbon atom is added, while adhering to the requirement for a planar arrange-

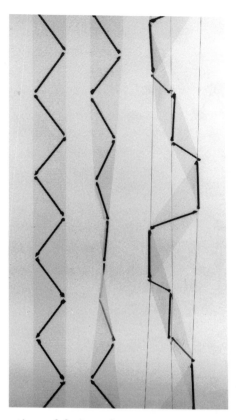

Figure 6.13.   Conformations of chain carbon atoms in crystalline vinyl-type polymers, exemplified by identical ribbons, where chain carbon atoms are alternately positioned at the opposing edges. (a) Flat ribbon as in polyethylene. (b) Slightly helical ribbon as in polytetrafluoroethylene. (c) Pronounced helical ribbon as in many of the polymers (e.g., isotactic polypropylene) being described next.

ment, two possibilities exist. One is akin to the trans position and the other to the cis position as described on p. 30 for molecules with a carbon double bond. Only the trans conformation can be repeated many times, resulting in a flat ribbon-like arrangement where chain carbon atoms are alternately positioned at the opposite edges of the flat ribbon. This structure can be found in polyethylene, nylon, and several crystalline vinyl polymers, and is commonly described as a planar zigzag conformation. The analogously flat cis conformation is not possible since after three repeats the available space would already be fully occupied.

Here another important aberration can lead to well-ordered structures that bear the possibility for crystalline arrangements. If the fourth chain carbon atom is placed right or left out of the plane (this positioning is generally termed + or − gauche), the new plane formed by the carbon atoms number two, three, and four will appear somewhat turned in the axis of chain growth.

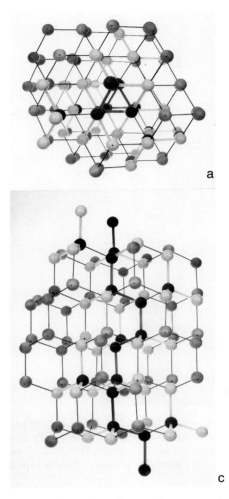

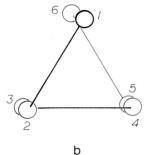

FIGURE 6.14. Isotactic polypropylene superimposed onto the diamond lattice [photos taken in sagittal (a) and vertical (c) direction of polymer chain]. (b) Consecutive arrangement of chain carbon atoms in isotactic polypropylene.

Depending on the sweep and the sequences of trans and gauche dispositions, the ribbon thus formed exhibits a more or less steep pitch. In Figure 6.13 three of these possible conformations are simulated by a picture of three identical ribbons consisting only of C—C bonds strung up at a 109° angle. One symbolizes the conformation of chain carbon atoms in polyethylene, another in polytetrafluoroethylene, and the third in isotactic polypropylene. In the third case, two triangular shapes positioned in the same plane are followed by two triangles turned 240° from the first plane. It is important to realize that these illustrations disregard those atoms that are extending from the chain, such as the methyl groups in polypropylene.

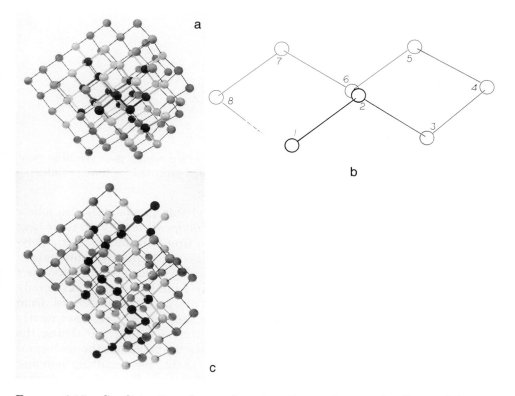

FIGURE 6.15. Syndiotactic polypropylene superimposed onto the diamond lattice [photos taken in sagittal (a) and vertical (c) direction of polymer chain]. (b) Consecutive arrangement of chain carbon atoms in syndiotactic polypropylene.

In another crystal modification of polypropylene the sequence of the chain carbon atoms follows a figure eight pattern. It is called syndiotactic polypropylene and here all the carbon atoms can also be located in the diamond lattice along a different direction as illustrated in Figure 6.15.

Since these chains occupy a greater volume than the zigzag chain in polyethylene, the specific gravity of polypropylene, 0.91, is lower than that of the similar linear polyethylene, 0.95.

Not all crystal conformations are so straightforward and clearly definable as the ones described to this point. The reason for this is that two different aspects must be considered for the formation of these types of polymer crystals. First, the polymer chain must be conformed so that those spirally shaped elements accommodate at their surfaces all the extending side groups to exhibit a regular pattern with a balanced surface area and volume relationship. Second, these irregularly pole-shaped structures must be able to be packed, stacked, or nested into a crystalline framework to provide best space utilization and best secondary bond efficacy. Probably the best exemplary model for it can be seen in the cellulose crystal, which is described in detail on pp. 294 to 301.

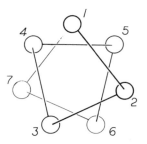

FIGURE 6.16.   Consecutive arrangement of chain carbon atoms in crystalline poly-methylpentene. The same helix arrangement has been proposed for fibrous sulfur.

The ethylene-type polymers with longer alkyl side groups will also follow these rules. Both *polybutylene (7)* and *poly-4-methylpentene (8)* with their long (and branched) side chains will crystallize. The first will, like isotactic polypropylene, form a helix with three monomer units (six chain carbon atoms) per turn, whereas the helix of poly-4-methylpentene will have seven monomer units per two turns as seen in Figure 6.16. Poly-4-methylpentene is the only optically transparent structural polyolefin. This peculiarity derives from the fact that both the crystalline as well as the amorphous form have nearly the same optical properties and specific gravities of 0.83 g/cm³ (being the lowest of all solid plastics). If one would endeavor to spread out the molecular structure even further by modifying the side chain, the resulting polymer structure would collapse, forming a lower melting plastic having a higher specific gravity and lower strength. Unlike polyethylene, in which—on crystallization—adjacent chains are able to closely approach each other, the polymethylpentene molecules occupy about the same volume whether arranged in a crystal or amorphous formation. At elevated temperatures the space requirement of the crystals is even greater.

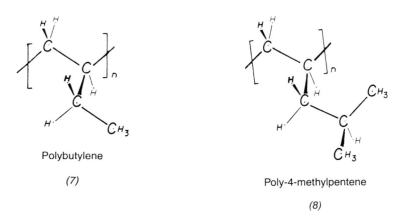

Polybutylene

(7)

Poly-4-methylpentene

(8)

Its high melting temperature of 450°F (233°C) can, in one respect be regarded as a drawback for this material since it necessitates molding conditions

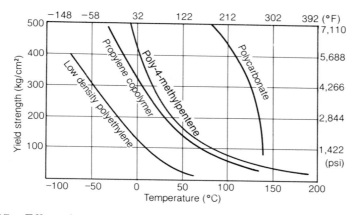

FIGURE 6.17.   Effect of temperature on tensile yield strength. Courtesy Mitsui Petro-
chemical Industries, Ltd.

close to its thermal stability range. Furthermore, the plastic is relatively brittle and susceptible to UV attack. Molded parts and films are clear and manifest the highest stiffness of all other polyolefins at high temperatures. Under
low load conditions, its stiffness may even surpass that of polycarbonate. The
curves shown in Figure 6.17 illustrate the great dependency of applied load for
retaining the shape of a part when temperature is increased. Other examples
are dealt with in Chapter 8.

Other vinyl-type polymers will not crystallize because of the presence of side

groups containing polar atom groups such as the $\diagdown C{=}O$ carbonyl group.

Some of them are *polyvinyl acetate (9)*, and all *polyacrylates (10)* and *polymethacrylates (11)*, but *polyacrylonitrile (12)*, being an exception in many properties, can crystallize.

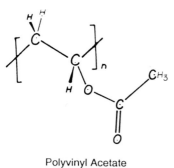

Polyvinyl Acetate

(9)

Polyacrylate

(10)

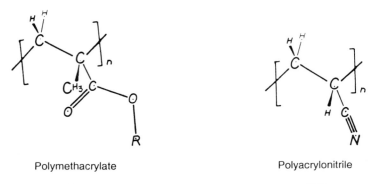

Polymethacrylate   Polyacrylonitrile

(11)   (12)

Polystyrene, the ethylene derivative with a pending phenyl group at the side
will not crystallize if it has been obtained by conventional peroxide polymerization. If styrene is reacted with special polymerization catalysts, ensuring a
stereo-regular positioning of the phenyl side group, it will crystallize readily in
a helix form with three monomer units per turn (same as the isotactic polypropylene, see Fig. 6.14). However, this isotactic polystyrene is of no commercial importance. Still the rigidity of the six carbon atom aromatic ring is being
passed on to the amorphous polymer, thus making polystyrene one of the
plastics with the highest modulus of elasticity under lowest deflection conditions. Therefore a polystyrene part dropped on a hard surface will sound,
unlike any other plastic, much like a metal part.

Polymer chains bearing atoms other than just carbon can also readily crystallize. Among these representatives are the *acetal polymers (13, 14)* also described as polyformaldehyde or polyoxymethylene.

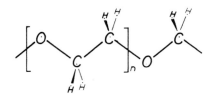

Acetal Homopolymer

(13)

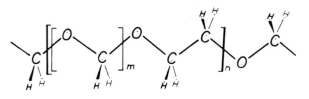

Acetal Copolymer

(14)

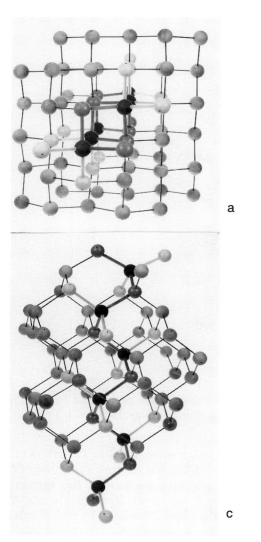

a

b

c

FIGURE 6.18. Chain conformation of polyoxymethylene (homopolymer acetal plastic) superimposed onto the diamond lattice [photos taken in sagittal (a) and vertical (c) direction of polymer chain]. (b) Consecutive arrangement of carbon and oxygen atoms in polyoxymethylene.

Even though this polymer consists of an equal number of —CH_2— and —O— links, the polymer chain atoms will fit into the lattice of diamond, though slightly constrained because of differences in angle and distance. In Figure 6.18 the acetal macromolecule is also superimposed onto the same diamond lattice utilized for the previous illustrations. The carbon atoms are shown in black, the oxygen atoms in gray, and hydrogen atoms in white. The carbon–oxygen chain bonds are highlighted.

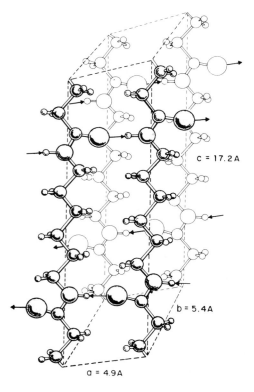

FIGURE 6.19. Nylon chain conformation. Reproduced with permission by Melvin I. Kohan, from Nylon Plastics, published by John Wiley & Sons, Inc., 1973.

The family of *polyamide* (nylon) resins, which consist of 4 to 12 carbon atom links with an interspersed nitrogen containing amide link *(15)*, results in especially high melting semicrystalline polymers because of the attraction between the parallel chains augmented by firm, strong hydrogen bonds. The chain conformation consists generally of a zigzag form as in polyethylene. Because of the presence of $\diagdown$C=O and $\diagdown$NH groups in the chain, a second, more gentle zigzag shape extends over the length of several monomer units (Fig. 6.19).

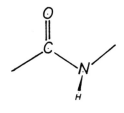

Polyamide

(15)

The formation of layers accommodating these hydrogen bonds contributes to the additional stiffness of these crystallites, but hinders the formation of equally high regularity in the crystallites when compared to polyethylene. These high forces provide sufficient strength to molded products even though the polymer has a lower molecular weight. Under molding conditions these hydrogen bonds are severed, resulting in the comparatively very low melt viscosity of nylon polymers. To a certain degree nylon resins behave like copolymers of ethylene and amides. The longer the —CH$_2$— chains are, the more they resemble polyethylene. Table 5.1 (p. 108) shows, for instance, how the crystal melting points decrease with increasing length of the —CH$_2$— segments.

Recalling the data given in Table 4.3 (p. 45) on three hydrocarbons of about equal molecular weight, all containing six carbon atoms, the great differences in their melting points must be recognized as impressive. Similar improvements should therefore be expected for all polymers containing *aromatic rings* as polymer chain members, substituting for some of the aliphatic chain members. The conformation of the simplest carbon chain molecule, paraffin, has been described explicitly, therefore the spatial arrangement of the simplest aromatic ring-containing compound, benzene or benzol, should be shown (Fig. 6.20).

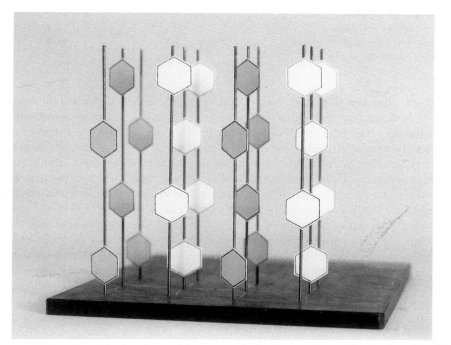

FIGURE 6.20.   Crystal structure of benzene. A similar conformation is found in high strength fibers containing benzene rings in their molecule. In those cases the aromatic rings become stretched out in the direction of the steel rods and somewhat tilted for better occupation of space.

TABLE 6.1. Crystallizing Propensity of Polymers with Aromatic Chain Links

Polymer	Glass Transition Temperature, T_g		Crystalline Melting Point, T_s	
	°F	°C	°F	°C
Noncrystallizing				
Polyarylate	375	190		
Polysulfone	375	190		
Polyetherimide	420	215		
Polyarylsulfone	430	220		
Polyethersulfone	450	230		
Polyamideimide	530	275		
Polyimides	590–690	310–365		
Provisionally crystallizing				
Polycarbonate	300	150	450	230
Polyethylene terephthalate	170	77	500	260
Copolyester of ethylene glycol and cyclohexanedimethanol terephthalic and isophthalic acid (PETG)	178	81	510	265
Prevailingly crystallizing				
Polybutylene terephthalate			480	250
Aromatic polyamide			530	275
Polyphenylene sulfide			555	290
Polyaryletherketone (PEEK)			635	334
Wholly aromatic polyester			780	415

In the model the steel rods were used only to position the benzene rings at the required spacing. In reality they are held in position only by weak van der Waals forces extending in all directions. In the case of many oriented or crystalline polymers, containing benzene rings, those rings happen to be arranged in a very similar fashion. In these cases the steel rods, as seen in the model, should be visualized as reflecting the other chain links that connect the rings via covalent chemical bonds. The spacing of the rings in the lengthwise direction will individually reflect the length of the chain, —C—NH— in the case of the aromatic polyamides and —C—O—CH$_2$—CH$_2$—O—C— in the case of one polyester. The benzene rings will become somewhat inclined to conform with the zigzag arrangement of the other chain members. In general, symmetrically structured chains and chains with a number of highly flexible —CH$_2$— links in a row will more likely tend to crystallize than the otherwise stiffer and unbalanced chains.

Aromatic rings will always impart a high rigidity to the chain. Although they may favor an association in many cases, they may also immobilize the chains, preventing proper alignment in a lattice. Therefore aromatic rings containing polymers exist both in semicrystalline and amorphous polymers.

Under all circumstances they will have higher glass transition temperatures, higher melting temperatures, and a higher modulus of elasticity if compared with polymers containing only aliphatic chain members.

Table 6.1 lists the melting temperatures of several polymers containing aromatic rings. The chemical structures of these polymers can be found at the end of Chapter 5.

LIQUID CRYSTAL PLASTICS

All crystalline materials are characterized by the existence of building blocks positioned in a three-dimensional lattice. If the ordered spacing in one of those three directions is discontinued, the forming substances lose their rigidity and their ability to resist mechanical forces. Substances of this kind are called liquid crystals by materials scientists and physicists since they behave in some respect like crystals but also retain their mobility similar to other liquids.

The peculiar electrical and optical properties of some of these materials have given liquid crystals their great popularity in liquid crystal electronic displays and the colorful thermal sensors. Other types of liquid crystalline materials are present in living tissue and are of importance in biological processes.

In contrast to the above listed liquid crystal materials, which reside predominantly in that mesogenic state, in liquid crystal polymers this condition exists only for a very brief time during melt or solution processing. Unlike regular crystal formation fostered by van der Waals attractive forces, the crystal formation in liquid crystal polymers is induced indirectly as a result of the rod-like or plate-like shape of at least parts of the polymer molecule. As a theoretical limit for the ratio of length to diameter of the rod-like sections, 6:3 has been established. The liquid crystal state could be compared with a box of randomly poured nails, which will align in parallel bunches to occupy less space if subjected to any random motion. Only within the narrow liquid crystalline state, the nails are not "glued" together so they can easily slide past each other but still prefer to retain their regional alignment.

As the tendency for this formation to occur diminishes as the aspect ratio decreases (down to stubby screws), so will many liquid crystal polymers need some help during processing to obtain that preferred order.

Just the chemistry of the nematic phase, which represents only one of the four liquid crystal phases, should be considered here, since it can bring about the most favorable processing conditions and the best mechanical properties. A typical nematic liquid crystal polymer system may consist of a compound in which two benzene rings are connected by two nonrotatable chain links so that both rings can become fixed in parallel plane positions (16).

These thermotropic liquid crystal compounds are characterized by two melting points which may be a few to 100°F (up to 50 K) apart. The lower temperature (T_{CN}) indicates the transition temperature from the crystalline to the nematic phase and the higher temperature (T_{NI}) the transition from the nematic phase to the isotropic melt as shown in Figure 6.21. Processing must occur

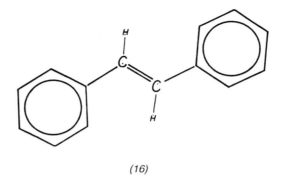

(16)

close to this narrow nematic temperature range, so that shear flow can be utilized to align the crystallites parallel to the flow direction. With nematic liquid crystal polymers, where the ordered crystalline parts consist of rigid or semirigid, stretched polymer chains, high strength and high modulus of elasticity materials are obtainable. This orientation will enhance properties in the flow direction of the molded part significantly, resembling those of fiber-reinforced plastics.

The very high melt temperature of some liquid crystal polymers provides some advantages for applications in which plastics are sought to replace stainless steel and ceramics in moderately high-temperature applications, but the increasing processing difficulties still allow room for the lower melting varieties that are now in commercial use. A few of the liquid crystal plastics can be processed only by powder pressing and sintering, since their melting temperature is too close to the decomposition temperature.

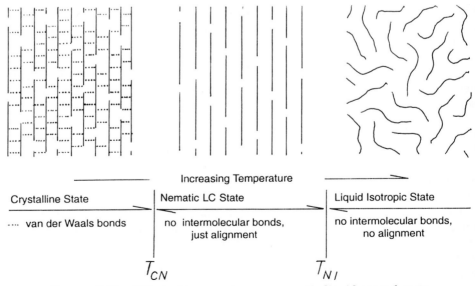

Figure 6.21.   Effect of temperature on nematic liquid crystal state.

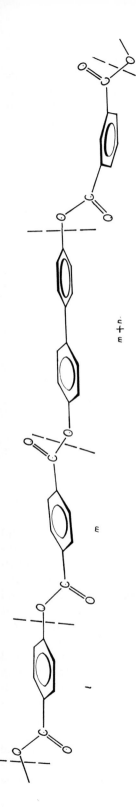

Aromatic Copolyester of *p*-hydroxybenzoic Acid, Biphenol, and Iso-
and Terephthalic Acid
655°F (346°C)

(17)

A small selection of the great number of liquid crystal polymers is now marketed as liquid crystal polyester plastics consisting of aromatic carboxylic acids and aromatic hydroxy compounds (see also Chapter 5, p. 114). Their exemplary chemical components with their approximate heat deflection temperatures at 264 psi (1.8 MPa) are shown in declining order in structures (17), (18), and (19).

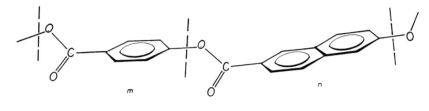

Aromatic Copolyester of *p*-hydroxybenzoic Acid
and 6-hydroxy-2-naphthoic Acid
445°F (230°C)

(18)

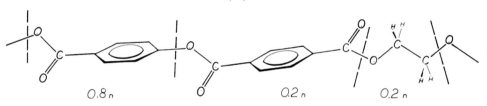

Aliphatic-aromatic Copolyester of *p*-hydroxybenzoic
Acid and Polyethylene Terephthalate
325°F (163°C)

(19)

Just as in other high-temperature thermoplastics, the chemical (hydrolysis) and flame resistance improves with increasing melting temperature and increasing crystalline content. Further advantages of these liquid crystal plastics are that their thermal stability and low melt viscosity allow fast production cycles with only minor losses in properties when reusing reground material. The easily obtainable orientation in the flow direction makes it possible to mold parts with very high strength and modulus in the flow direction. On the negative side, this high orientation ratio can lead to difficulties, including insufficient strength in the cross directions, especially along weld lines, and easy surface abrasion (fibrillation). Inconsistent orientation may also cause part warpage. Specific solutions for these problems are being developed selectively by keeping flow direction under control, by distinct positioning of gates, and by push-pull flow (flushing material repeatedly through the cavity from oppositely positioned gates).

Inasmuch or while it is common to reduce crystallinity in semicrystalline thermoplastics by various means, so liquid crystal polymer blends and composites will become established for producing parts with molecular composites in a self-reinforcing, three-dimensional, microfibrillar structure.

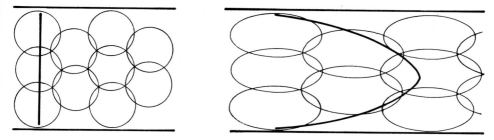

FIGURE 6.22.   Schematic sketch illustrating the slight orientation obtained under any processing involving material flow.

ORIENTED POLYMERIC MATERIALS

Whenever a melt consisting of intertwined long chain molecules is subject to flow, a speed differential will exist between the outer layers close to a stationary surface and the center part as a result of the viscosity (equal to internal friction) of the melt. This shear force will align at least part of the chain links in the flow direction. A randomly spread out molecule positioned within a certain sphere will assume the shape of an ellipsoid (Fig. 6.22). Since most polymers display dichroism (which means having different absorption coefficients for light polarized in the length and cross directions) even this small amount of orientation can easily be identified by placing a flat sample between two crossed polarizing light filters and thus identifying the direction and the magnitude of orientation.

All injection-molded and extruded parts will display this phenomenon to a certain degree. In practical terms this mild orientation will neither affect nor improve mechanical properties or the serviceability of molded parts. However, under extreme conditions (e.g., heating the part beyond the heat deflection temperature), the part may display excessive shrinkage or distortion.

Particularly during extrusion processing, additional stretching of the plastic melt will occur during draw-down. Since that orientation also occurs at a relatively high temperature, the pulling force is usually low because of the low viscosity of the exiting plastic, allowing adjacent molecule sections to slide easily by each other. To avoid the possibility of experiencing disturbing shrinkages, especially during high-temperature usage or processing (thermoforming), the extent of orientation should be monitored and kept under certain limits. The magnitude of this orientation will seldom exceed 10%, as can be readily determined by heating a square piece of a film or sheet in an oven just below the melting range for a short time and measuring the dimensional changes occurring in that piece. Even high rates of draw-down (50% or more) contribute little to any beneficial orientation but have more effect on the dimensional stability of extruded profiles.

A significant chain segment orientation may be achieved when high shear forces are applied at lower temperatures, which are closer to the glass transition temperature. Under these circumstances additional polymer chain segments of the originally randomly coiled molecules will be aligned in the direction of pull, leaving fewer tie-chain segments remaining in the transverse

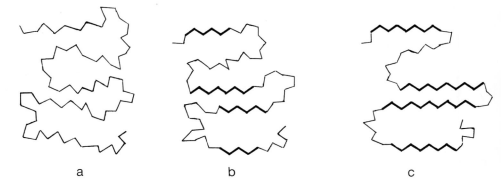

a   b   c

FIGURE 6.23.   Schematic positioning of chain segments in extrusions. (a) Randomly coiled chain segments, (b) melt flow oriented segments preferentially tilted in the direction of pull (about 10% extension), and (c) high strain oriented segments locked in parallel to the orientation direction.

direction. Since the polymer chains do not consist of smooth tubular structures but of repeatedly narrowing and widening sections, parallel alignment of long stretches should be assumed to be favored by the jerking motion exerting additional forces onto unaligned sections. Since extensions may range from 200 to 1000% (2- to 10-fold pull), the amount of work (force times distance) will reach appreciable levels. A schematic positioning of chain segments is shown in Figure 6.23.

Usually a heat setting process follows the stretching to stabilize the oriented product so that further dimensional changes will not occur during use when the product may become subject to elevated temperatures.

The orientation process, though accomplished under widely varying conditions, should first be illustrated on the basis of a stress–strain diagram, obtained on a strong, ductile plastic tensile specimen (Fig. 6.24).

When a tensile force is first applied to a thin (about 1/16 in. (2 mm) thick) specimen, the force to be exerted will rise rapidly but cause only a minor extension (about 1% in length) of the part. If the force is removed, the specimen returns to its original length, since aside from a slight elastic twisting of the C—C—C chain angle (much rather than a stretching of the C—C covalent bond) no relative displacement between chain links has occurred. Under further stretching, greater extensions will take place but the force will increase only up to a certain point, the so-called yield strength where a viscoelastic flow will become evident by a successive plastic flow, extending the specimen 5 to 10%. Only about 1 to 3% of this extension can be recovered immediately on removal of the tensile force. If stretching is continued, a sharp indentation (neck) may form at some place along the specimen in certain materials. Others will also become reduced in cross section but in a balanced way. This neck will spread in both directions on continued stretching until the full length of the specimen has been drawn out at a force nearly remaining constant. A further stretching may impart strain hardening of the sample (increasing the force)

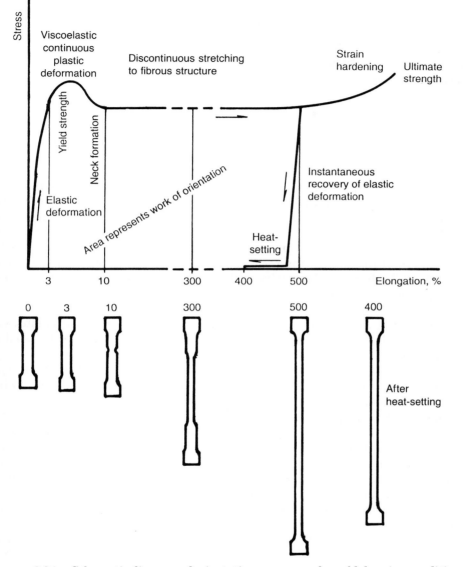

FIGURE 6.24.   Schematic diagram of orientation process under cold drawing conditions and with heat stabilization.

but will eventually lead to fracture at the point of ultimate tensile strength. It should be noted here that the true tensile strength at break is considerably higher than the force value indicates since the actual cross section of the specimen just prior to failure has become much smaller.

For orientation processes, temperatures are higher than room temperature and the stretching process ends much sooner. On removal of the tensile force only about 1 to 3% of the elongation can be recovered instantaneously. During the following heat-setting process either an additional amount of the viscoelastic extension will become relaxed or under retention of the gained dimensions just part of the first imposed viscoelastic stresses relieved. The bulk of the drawing work remains locked in by the van der Waals forces acting between stretched out chain links.

To retain dimensional stability, the use temperature range must

1. be kept below the heat-setting temperature, or
2. remain below the glass transition temperature of the plastic, or
3. the intermolecular bonds must be strengthened by crystallite formation.

Since it is not possible to achieve orientation of polymer chains inside of crystallites, semicrystalline polymers are usually first melted at high temperatures then quenched and reheated to the desired drawing temperature.

The properties of the materials subjected to that kind of orientation are notably different from those of the starting material. Because of the parallel orientation of the chains, the resulting material will be more tightly packed, will have a higher specific gravity, and will be less soluble in solvents and less susceptible to environmental stress-cracking.

Although orientation (like crystallization) will not extend to 100% of the bulk material, the boundaries between oriented and nonoriented regions will be so subtle that the differences in optical properties will not be manifested in an opaque appearance as is the case with partially crystalline plastics. As in cross-linked polymers, in oriented polymers the extent or likelihood for crystallization diminishes since the polymer chains are already tightly connected and are not that free to align properly in a crystal lattice.

However, the single, most important reason for orienting polymeric materials is the improvement attainable in mechanical properties. On a molecular level, tensile properties are always higher in the direction of the covalently bonded carbon–carbon chain than in the transverse direction dominated by the much weaker van der Waals bonds. Therefore unidirectionally oriented polymers may have tensile strength values 10-fold higher than the strength of nonoriented polymers. This is also true for the modulus of elasticity, though not at such a magnitude (Table 6.2).

The gains obtained in the length direction are offset by losses in the transverse direction because of the reduction of valence bonds positioned in that direction to a few tie bonds. A highly oriented tape can therefore easily be split

TABLE 6.2.   Properties of Nylon 6/6

	Tensile Strength		Modulus of Elasticity		Elongation (%)
	ksi	MPa	ksi	MPa	
Molded part	11	75	300	2000	250
Oriented fiber	60–140	400–1000	1000–1700	7000–12000	30

into separate strands of parallel fibers. Since an appreciable portion of the material's potential elongation has been consumed during the drawing process, elongation values of highly drawn fibers are substantially lower. In cases of biaxially oriented films this may not be true because many of these materials (e.g., polystyrene) might start out as a low elongation, brittle film.

Unidirectional orientation processes are ideally suited for the production of fibers. Since only a tensile force is involved in fiber drawing, the fiber cross section can be easily reduced undisturbed by transverse forces. Under similar drawing conditions, massive parts would soon fail because of fracture (see Chapter 7). All important synthetic polymeric monofilaments, staple fibers, or bristles include a cold drawing process in their production.

In synthetic polymeric film and sheet products the situation is more complicated. Some of those products are solvent cast, cast as catalyzed monomers or cast on chill rolls in extrusion processes, all resulting in negligible orientation. Blown film, extruded thin sheet, or calendered sheet products represent in-melt-formed products exhibiting only moderate orientation of the polymer chains. Analogous to fiber drawing, true orientation is obtainable only by the application of high tensile forces at relatively low temperatures. In most cases the longitudinal extension is augmented by a simultaneous or consecutive extension in a transverse direction. Exclusively or dominantly unidirectionally oriented films are prevailingly used for heat shrink applications. The material selection and the processing parameters for them will determine the extent of shrinkage, the temperature at which shrinkage will occur, and the orientation release stress exerted during shrinkage. In most of these cases orientation will take place at lower temperature than is practical for orientation of conventional films. Some examples are high-shrink polyethylene terephthalate films, polyolefin, and polyvinyl and polyvinylidene chloride films.

Biaxial orientation contributes the most beneficial improvements to film products, such as tensile strength and especially dart impact strength and fold endurance.

Barrier properties will depend on mutual chemical affinity. In cases in which high permeability through nonoriented material depended mainly on "free volume" (void content) an improvement will be noted. In most cases reduced gas transmission rates will depend more on crystallinity and geometry than on orientation.

Products such as biaxially oriented polypropylene and polystyrene have gained importance in the packaging field. It is remarkable that both films are quite tough and extendable even though they are very brittle (polypropylene at low temperatures only) prior to orientation.

Highly sophisticated orientation processes have resulted in the extremely dimensional stable and solvent-resistant polyethylene terephthalate films. Among the mass produced articles, the injection-blow molded carbonated beverage bottles should be mentioned here. The biaxial orientation during stretch-blow molding of the heated injection molded polyethylene terephthalate preform is responsible for their excellent properties. In Figure 6.3 the changes in the shape of that part during processing are exemplified. The right-hand side shows the achievable regression of this orientation by heating to 500°F (250°C) and the white coloring of the nonoriented neck resulting from heat-promoted crystallization.

Regardless of the orientation or forming process employed, the oriented products can be nearly returned to their original shape by heating the part to the molding or melting temperature. To eliminate gravitational forces, heating can be done in silicone oil. Even an injection molded part can, under these conditions, be returned to a slug of plastic. This phenomenon, called plastic memory, is proof that in spite of all the sliding and displacing of chain links, a small but significant number of interconnections will be retained, which, on the melting of the plastic, will constitute the centers around which the mobilized chains will again regroup. More of this phenomenon is described in the paragraphs on elastomers. This behavior is unique to plastics and is not shown by metals, ceramics, or glass, since only plastics can consist of long intertwined chains.

A few specific examples of materials should be given that pertain to orientation processes:

A somewhat cross-linked polyethylene or polyvinyl chloride tubing can be expanded three or more times its original diameter and cooled right away. The orientation in that heat-shrink tubing can be reversed by heating it above the temperature at which it was extended, thereby tightly clasping any oddly shaped object to be enveloped.

Biaxially oriented polystyrene film and sheet drastically alter their properties from the nonoriented form. They become puncture and tear resistant and will not fracture when folded or creased. Their surfaces become harder and more scratch resistant.

Polymethyl methacrylate sheets, also an amorphous material, can be thermoformed without affecting its properties. It is the plastic that best remembers its prior shape, since formed parts easily return to a smooth flat sheet when reheated. However, polymethyl methacrylate sheets, when used for aircraft and security glazing applications, must be treated differently to increase their strength. Forming takes place at lower temperature under great mechanical forces to obtain true biaxial orientation. By proprietary processes and surface treatments the high internal stresses are balanced to prevent environmental stress-cracking or crazing.

The behavior of the thermoplastic polyesters is more complex since many of them also exist in highly crystalline forms. The most important of these polyesters, polyethylene terephthalate, was, as originally produced, unsuitable for injection molding since crystallization was uncontrollable and molded parts— once spontaneously crystallized—became extremely brittle. The same resin, however, when extruded into fibers or films and properly oriented under controlled temperature conditions, became the well-established polyester fibers and films. A heat-setting process had to be incorporated to prevent shrinkage of fiber garments during ironing and any minute changes of the graphic, photographic, and electronic films during processing and storage. Other films of the same material to be used as heat-shrinkable films were of course not stabilized. Once oriented, further crystallization is prevented. Both orientation and increasing the molecular weight will hinder the growth of crystallites. The small size and low concentration of the remaining crystalline matter give polyethylene terephthalate films their high clarity.

Later, polyethylene terephthalate with about twice the high molecular weight (80,000) was formulated by incorporation of nucleating agents so that a rapid controlled crystallization can be achieved during the molding process. As a result of this change much tougher parts with good rigidity similar to polybutylene terephthalate were obtained. Polyethylene terephthalate compounds containing one of the patented nucleating agents that speed up crystallization can restrict crystallinity to about 30%. This results in a product combining several optimal properties:

1. higher rigidity at elevated temperature and better thermal resistance and
2. toughness even at low temperature and good barrier properties.

This best available, oven usable food-packaging material is suitable for heating food in either a microwave or a conventional oven.

The high temperature rigidity of the crystallized polyester makes it impossible to thermoform an already crystallized sheet or film. Thermoforming the noncrystallized material into a desired shape by conventional methods and then attempting to crystallize the formed part by heating would grossly distort its shape and would require a long processing time. Therefore the forming process must be combined with the crystallization step. Figure 6.25 illustrates that the fastest crystallization occurs between 300 and 360°F (150 and 180°C). Crystallization starts when the sheet is heated in the oven but must be continued in a heated mold at 250 to 300°F (120 to 150°C) for 2 to 6 sec. The progress of crystallization can be followed by observing the onset and intensity of the opaqueness of a clear (unpigmented) sheet material. Since the properties of the formed part are very dependent on the extent of crystallinity, the establishment of its value should become a part of quality control. The percent crystallinity can be absolutely established by determining the polyester's density. This nearly linear relationship indicates a crystallinity of 25% at a density at 77°F (25°C) of 1.360 g/cm^3 and 50% at 1.390 g/cm^3.

In other polyesters ethylene glycol is (partially) replaced by cyclohexanedimethanol, also a glycol, to prevent crystallization. The introduction of this second six-membered ring, even though zigzag shaped, into this polymer secures sufficient structural integrity to these parts, without the need to orient or crystallize them.

To stabilize highly oriented plastics, one of two conditions must be fulfilled: either (1) the glass transition temperature must be higher than the temperature of use, or (2) the orientation must be fixed by the formation of crystalline regions. Many of the useful oriented film materials belong to the latter group, including polyolefins and polyvinyl fluoride.

Although the extension, when carried out to 3 to 10 times the original length achieves a remarkable alignment of the majority of chain links in the direction of applied force, it still contains too many weak, obliquely positioned chain segments. To use polyethylene as an example, only when practically all chain

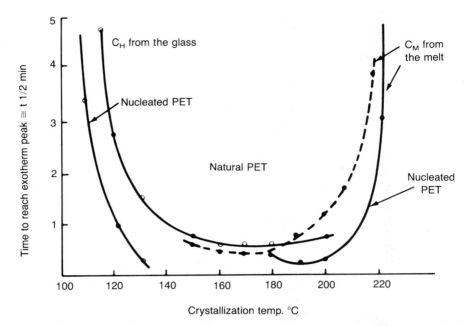

FIGURE 6.25. Crystallization half-time of natural and nucleated polyethylene terephthalate. The inverse crystallization rates for PET are shown on the left-hand side as they are observed when heating a sheet. The crystallization rates observed when cooling a melt of this polymer are shown on the right-hand side. Copyright Eastman Kodak Company. Reprinted by permission.

links are aligned along a straight zigzag line can a pattern resembling the lonsdaleite crystal lattice shown in Figure 6.2 be obtained, thus forming a material with a tensile strength of 200,000 psi (1.5 GPa) and a modulus of elasticity of 15 million psi (100 GPa). This microfibrillar structure can be secured by a tedious gel-spinning process out of ultrahigh-molecular-weight polyethylene involving a 100-fold extension in the pull direction. This process is also called superdrawing. The highest strength values reported are above 1 million psi (above 6 GPa).

Figure 6.26 shows how a remarkable improvement can be obtained in aligning chain segments in the pull direction when a 100-fold extension is employed instead of only a 5-fold extension. Half of all originally randomly positioned chain segments will end up with 0° deviation from the fiber direction.

In the case of the poly(p-phenyleneterephthalamide) fiber, which in addition possesses high temperature applicability, extensive tugging and pulling are not required since this liquid crystal polymer aligns itself during the lyotropic, liquid crystal solution spinning process. Another process adapted from the metallurgical reverse extrusion process utilizes solid state or hydrostatic extrusion to obtain outstanding mechanical properties.

At this point it should be briefly indicated how other structural materials achieve their peak tensile properties.

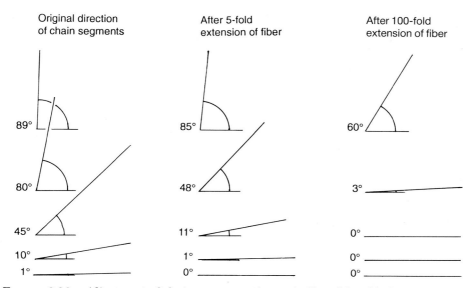

FIGURE 6.26.   Alignment of chain segments theoretically achievable by a 5- and 100-fold extension.

The tensile strength of metals can be enhanced by drawing, but most of the applied force is extended for the work connected with the reduction in cross section and only a minor part will be expressed through the directional extension of the shape of grains. The ultimate tensile strength is obtained with acicular metal whiskers, which consist of single crystals with exceptional length-to-diameter ratios. The expensive processing steps prohibit their broad application.

In minerals the exceptional properties of fibrillar or lamellar materials are less based on the improvement of properties in the fiber or planar directions but more on the introduction of looser bonds in their transverse directions.

Ceramic and glass materials in general show no directional preferences in properties. Even the glass that is widely used in glass fibers for reinforcing plastics has no higher modulus and strength in the fiber direction than in any of the other directions. That is one of the reasons why glass fiber-reinforced plastics do not suffer much under shear and transverse impact conditions as is the case for the highly oriented fibers including carbon and graphite. Ceramic whiskers behave in a manner similar to the metallic whiskers described above.

Among the natural structural materials both cellulose and protein fibers are dominant. The cellulose macromolecule consists of a series of six-membered rings interconnected by oxygen acetal linkages. Both the formation of this macromolecule and the way orientation is achieved remain a mystery. It is established that the monomeric two glucose molecules-containing units are oriented for optimum mechanical properties.

The advantage indigenous to all natural systems is that the synthesis of these materials coincides with the shaping and forming of the objects, which

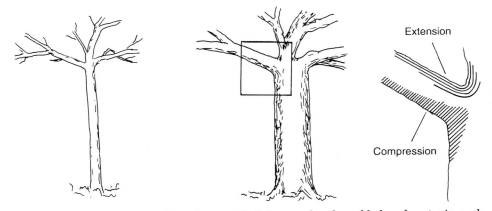

FIGURE 6.27. Young and old maple tree. Cellulose molecules added under strain each year.

are guided by rules that are not achievable. For most synthetic products the formation of the material and the shaping of the end product are processes usually not interconnected. In addition, no explanations have been found for the ability of living organisms to incorporate structural components under stress. As shown in Figure 6.27 the side branch of a young maple tree will be under tensile and compressive stress. If additional cellulose molecules would just be added around it, the increased weight would bend the branch downward in time. Additional layers must be added at appreciable levels of strain to maintain that desirable inclination.

The overall molecular structure (orientation and crystallization) of the native cellulose resembles very much those synthetic polymers that have gained their strength by orientation or oriented crystallization. Regenerated natural fibers such as cellulose-based rayon fibers are generally quite inferior to natural cotton fiber.

The natural protein fibers are also characterized by their highly ordered molecular and submicroscopic structure, in many cases parallel spiraled fibers. The high clustering of hydrogen bonds connecting the amino acid chains replaces in importance the crystalline ordering in cellulose. Nylon, chemically the closest relative among synthetic fibers, can obtain and also surpass the properties of natural wool and silk protein fibers. Without an orientation process this would not be possible.

Most of these natural products will be reexamined in Chapter 13 on composites.

REFERENCES

W. Brostow, Science of Materials. John Wiley, New York, 1979.

A. Ciferri, W.R. Kriegbaum, and R.B. Meyer, eds. Polymer Liquid Crystals. Academic Press, New York, 1982.

P. Corradini, Molecular dimensions. In Applied Fibre Science, F. Happey, ed., Vol. 1, p. 325. Academic Press, New York, 1979.

R.C. Evens, An Introduction to Crystal Chemistry. Cambridge University Press, Cambridge, 1964.

M. Grayson, ed. Encyclopedia of Glass, Ceramics, Clay and Cement. John Wiley, New York, 1985.

G. Gruenwald, Proposed structures and models for polymers crystallized in bulk. Journal of Polymer Science, 61, 381–402, 1962.

C. Hall, Polymer Materials, an Introduction for Technologists and Scientists, 2nd ed. John Wiley, New York, 1989.

H.D. Keith, Crystallization of long-chain polymers. In D. Fox, M.M. Labes, and A. Weissberger, eds., Physics and Chemistry of the Organic Solid State, John Wiley, New York, 1963.

A.I. Kitaigorodskii, Organic Chemical Crystallography. Consultants Bureau, New York, 1961.

S.L. Kwolek, P.W. Morgan, and J.R. Schaefgen, Liquid crystalline polymers. In Polymers: An Encyclopedic Sourcebook of Engineering Properties. J.I. Kroschwitz, ed., p. 509. John Wiley, New York, 1987.

N.R. Legge, G. Holden, and H.E. Schroeder, eds. Thermoplastic Elastomers—A Comprehensive Review. Hanser, Munich, 1988.

C.P. MacDermott, Selecting Thermoplastics for Engineering Applications. Marcel Dekker, New York, 1984.

B. Mandelbrot, The Fractal Geometry of Nature. W.H. Freeman, San Francisco, 1982.

L. Mandelkern, The crystalline state. In Physical Properties of Polymers, J.E. Mark, ed. American Chemical Society, Washington, D.C., 1984.

J.E. Mark, The rubber elastic state. In Physical Properties of Polymers, J.E. Mark, ed. American Chemical Society, Washington, D.C., 1984.

H. Tadokoro, Structure of Crystalline Polymers. John Wiley, New York, 1979.

B.M. Walker and C.P. Rader, eds. Handbook of Thermoplastic Elastomers, 2nd ed. Van Nostrand Reinhold, New York, 1988.

B. Wunderlich, Macromolecular Physics, Vol. 1, Crystal Structure, Morphology, Defects. Academic Press, New York, 1973.

7

Mechanical Properties of Polymeric Materials

EFFECT OF SHAPE AND STRUCTURE ON MATERIAL PROPERTIES

Since plastics are primarily used for structural applications, the mechanical properties of these materials are of great importance. However, use cannot be determined only by strength properties. Quite dissimilar materials with diverse strengths have proven to be satisfactory in identical application (e.g., wood, bricks, and steel are all used in building construction).

Another very important consideration for judging a material is to take its shape, the *scale effect*,[1] and its manufacture into account, since materials may have quite different properties in their bulk form than in distinctly structured forms. This has already been expressed in Table 1.2. The structural differences between heavy cast iron parts and thin cold rolled sheet steel parts are quite obvious. And the same cellulose macromolecule can impart softness to cotton or confer high rigidity to hardwood. These differences can also be found among plastics. In addition to geometry, in the case of polyethylene terephthalate the tensile strength values will change from molding compounds to biaxially oriented films and textile fibers (see also p. 174). Glass may find no application in bulk forms for any machinery part because of its susceptibility to brittle failure. However, used as reinforcement in fiber form, the same material becomes exceptionally useful. The glass's brittleness in bulk has been changed to high flexibility just by varying its dimensional proportions. Also vitreous enamels still retain their niche as surface coatings.

These quite drastic differences may arise even with much more subtle variations. Although the modulus of elasticity of nylon is quite modest, nylon

---

[1]The reader should recall the controversy over whether it is possible to reconstruct an insect to one thousand times its size just by multiplying every dimension of the insect by one thousand, since according to the "square-cube law" the replica's sheer weight would cause it to collapse. Even though the assumption that when upscaling a part the cross-sectional area, which has load-carrying capability, will increase as the square, and the weight of the structure will increase as the cube of its dimension is correct, there are other effects, such as the moment of inertia, that will counteract this. Still, certain size effects will remain.

parts—when small and compact—can sustain appreciable loadings and will be perceived as a very rigid material. In an application such as in washing machine hot water valves, there will be no danger of deformation. On the other hand, the same nylon resin as nylon tubing is not perceived as rigid but rather as easily bendable, ensuring applications in areas where persistent flexibility is required.

Furthermore, all mechanical properties are based on test specimens occupying a certain volume. Therefore, if comparisons are made between materials of quite different densities, the obtained strength values may be divided by their specific gravities to obtain comparable values for equal weight structures. These properties are usually designated as *specific strength* values. This will alleviate the discrepancy between the properties of metals and the properties of plastics. More on this subject, as well as tabulated values for various materials, will be found in Chapter 13. In applications in which occupied space and not weight poses design limitations, however, this consideration does not apply.

Already normal ambient conditions are subject to wide variations in regard to *temperature* and *humidity*. For many applications, around heat-generating equipment or where danger to exposure to certain chemicals exists, additional environmental conditions must be taken into account.

The *effect of time* will—much the same as temperature—influence properties. In this case three conditions have to be taken into account: the rate of load application, the duration of applied load, and the frequency for cyclic load applications. Other effects are caused by the combination of elevated temperature and extended time. In addition to creep, property changes may emerge as a result of chemical aging or stabilization of the metastable state of molded thermoplastic parts (in the literature also designated as physical aging).

All these influences will be covered in more depth in the following chapters of this book.

If all these variables are taken into consideration an unmanageable amount of data would be obtained for each mechanical property, such as tensile strength. Therefore generally listed property values are obtained not only under standardized and controlled environmental conditions but also under rigidly regulated test procedures on samples of narrowly defined dimensions. For this reason, values are quite reproducible and excellently suited to establish quality changes within a certain material.

However, if two chemically different materials are compared or if substantial variations exist in geometry, environment, or temperature, caution must be exercised in making comparative judgments as well as in merely considering the standardized published values. In many cases these reflect the optimum values obtainable and may deviate a great deal from parts molded under production conditions that may, in addition, include the utilization of compounds containing colored and reground materials.

MODULUS OF ELASTICITY

The next question to be considered is how to determine which of the various strength properties will reveal the most pertinent information. Tensile properties have a certain universal appeal. They are best related to the basic property describing the atomic bond strength. They are also the most significant property for fibrous materials. However, for structural applications, the flexural properties (a combination of material responses to tensile and compressive forces) are of overwhelming importance. Bending and buckling rather than the possibility of it ever being pulled apart will ultimately terminate the usefulness of any structural entity.

In considering the real requirement for a structural material or design, it is apparent that *excessive deflection, vibration,* or *flimsiness* will in nearly all cases pose limits very much sooner than breaking or crushing. This leads to the conclusion that the *modulus of elasticity,* which represents the material's stiffness, is the most important property.

Keeping all these admonitions in mind, it appears that the selection of a certain number for the modulus of elasticity in flexure [e.g., a minimum of 100,000 psi (1000 MPa)], will best classify structural material candidates for further consideration. For structures extending more than about 5 ft (2 m) a significantly higher modulus material must be chosen.

To defend the selection of this number it should be pointed out that an extension at the 1% level quite universally is not only the limit any structural contrivance might endure while still remaining serviceable, but also represents, based on the extensibility of chemical bonds in these materials, the limit up to which no permanent changes in the material will occur. This also is germane for long-term exposures to mild environmental influences. The higher elastic elongation values, about 5%, often found in thermoplastics are based on their quasielastomeric properties but cannot be relied on for repeated or long-term load applications.

The 1% extension value will mathematically translate into a minimum design stress level of 1000 psi (10 MPa). Indeed, these derived values are in most cases in agreement with the recommended safe stress limits published by manufacturers of engineering thermoplastics. These values are largely based on extensive testing and therefore provide a much more accurate value for the upper design stress limit than does dividing the published, highly inflated tensile strength values by an arbitrary safety factor.

As will later become apparent, aberrations between contemplated and listed values for the modulus of elasticity in flexure for alternate materials will be dwarfed by the likelihood that because of environmental or other circumstantial changes a potential foundering in other idealized high mechanical properties (tensile and impact strength) may occur. Once the primary selection step has been taken, minor variations in modulus of elasticity values of otherwise similar materials could still justify adjustments in cross section or other design aspects. Materials not passing the modulus of elasticity requirements can safely be eliminated from further considerations.

Since the modulus of elasticity is such an important mechanical property for structural parts and since this property is so reliable for its use for engineering design calculations, the ranking of several important plastics and other competing materials according to modulus of elasticity is given in Table 7.1. The modulus of elasticity can be greatly increased by fibrous reinforcements. In most cases the kind and amount of reinforcing fibers and especially their

TABLE 7.1. Modulus of Elasticity in Flexure of Unfilled Thermoplastics and Other Structural Materials

Materials	Modulus of Elasticity in Flexure	
	ksi	MPa
Polyethylene, low density	25–50	175–350
Polybutylene	50	350
Fluorinated ethylene–propylene polymer	90	600
Cellulosics	100–1000	700–7000
Nylon polyamides	140–400	1000–2750
High-impact polystyrene	150–400	1050–2750
Polyethylene, high density, polypropylene, polymethylpentene	200	1400
Acrylonitrile–butadiene–styrene polymer	200–400	1400–2800
Polyvinyl chloride, rigid	300–500	2000–3500
Polyphenylene oxide blends	325–400	2200–2800
Polycarbonate, polybutylene terephthalate, polysulfone	350	2400
Polyethylene terephthalate, polymethyl methacrylate	350–450	2400–3000
Acetal polymers	380–430	2600–3000
Polystyrene	450–500	3000–3500
Polyetherimide, polyimide (thermoplastic)	480	3300
Polyphenylene sulfide, polyacrylonitrile	550	3750
Polyetheretherketone	560	3800
Polyamide-imide	700	4800
Phenolics woodflour, cotton filled	850–1400	6000–10,000
Glass fiber-reinforced thermoplastics	1200	8000
Wood soft and hard	1600	11,000
Bone (animal), concrete	3000	20,000
Ceramics, bricks	4200	30,000
Magnesium	5700–6500	39,000–45,000
Glass	7000–11,000	48,000–75,000
Glass filament wound composites, aluminum, fused silica	10,000	70,000
Gray cast iron	14,000	100,000
Copper, hard	17,750	120,000
Ductile cast iron	24,000	165,000
Stainless steel	28,000	195,000
Carbon steel	29,000–30,000	200,000–210,000
Diamond	163,000	1,100,000

orientation will influence values to a much greater extent than those based on the plastic itself. Therefore, only the properties of noncompounded, neat plastics will be considered here. Still, some—perhaps unfair—comparisons must be made when quoting some other conventional materials. When elastomeric plastics are excluded, one realizes that the variances among structural plastics are really minor in contrast with those of the other listed materials. This again indicates that the properties of plastics are dominated by the strength of the covalent carbon–carbon bond, the common feature of all listed plastics.

TENSILE PROPERTIES

Properties determined under *flexural loading* most closely duplicate stresses experienced under practical structural applications. But because of complexities, when the determination of flexural properties is extended beyond the elastic limit, not much can be learned from those tests about specific materials. In the case of *tensile tests,* the stress conditions are much more elementary and truly unambiguous, at least up to the elastic limit. To promote and prolong this condition, a tensile specimen usually consists of a long central column to which all permutations during the test should be limited. They gradually expand into two much heavier end pieces, which must remain unaffected and serve only to firmly grip the specimen in the testing apparatus.

Usually the specimen is extended at a predetermined, constant strain rate and the resulting stress (load acting at the upper jaw) is determined. Both the load in pounds (kg) and the deflection in inches (cm) over a fixed gage length can easily be recorded on graph paper. These are then converted to stress and strain by calculations using the original section dimensions and the increase in length. The slope of the initial straight portion of the curve represents the modulus of elasticity in tension. The numerical values are obtained as the ratio of increment of stress, psi (kPa), to the increment of strain, inch per inch (cm/cm). The shape of one such curve has been explained in Chapter 6 (p. 173).

Only the first part of that curve represents a straight line, and if that line would be extended to 100% elongation, the pertinent stress level would give the modulus of elasticity in tension. Only within this short straight section of the curve do all deformations remain truly elastic, which means that on removal of the load the original shape (length) of the specimen will be instantaneously reestablished. As mentioned, this stretch extends to about 1% elongation and represents the working limit for most structural materials. It should be noted that in metals the limits may be less than one-tenth that value. Up to this point—also on the atomic level—the extension will be uniform in the direction of pull without involving any sliding or transverse motions. This straight line becomes slightly bent with many thermoplastics and may extend to about 5% elongation.

Another conspicuous difference between metals and plastics must be considered. In metals the recoverable strain cannot exceed the 1% limit under any condition, the same amount as determined for the elastic limit. This also pertains to steel springs, which only appear to have high material exten-

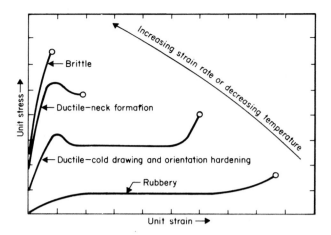

FIGURE 7.1   Typical stress–strain curves for thermoplastic materials. From J.R. Fried, Plastics Engineering, 33, July 1982.

sibility. In plastics the strain recovery processes are more complex and are known to reach 500% under certain elevated temperature conditions, which are akin to those of elastomers, in spite of the apparent 5% limit established under room temperature tensile test conditions.

Figure 7.1 shows a number of graphs that could have been obtained from a variety of plastic materials. Materials with the steepest starting slope would be representative of stiff engineering plastics and the plots with shallow slopes of elastomeric and rubbery materials. This initial slope is a very specific material constant that is alterable *only* by changing the test temperature, but is independent of the geometry of the specimen, the rate of load application, or the environment. Since the modulus of elasticity is such a reliable number, its use for design purposes has many advantages.

To understand the complications arising in the progressions of the curves beyond the initial, small extension region, the entailing phenomena on the atomic level must first be understood. The initial extension detected is composed of an increase in the distance between atoms and an opening of the bond angle (Fig. 7.2). Even in highly oriented and crystalline materials (including polymers) the majority of bonds will be positioned at a diagonal angle rather than straight in the direction of pull.

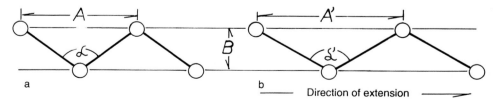

FIGURE 7.2.   Positions of atoms (atom groups) (a) at rest and (b) under elastic strain (exaggerated). Initial changes take place in distance A and angle α but not in distance B.

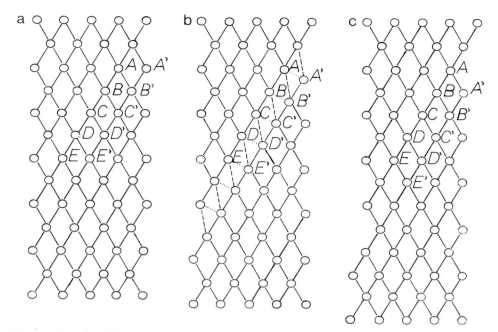

FIGURE 7.3. Positions of metal atoms (a) prior to extension, (b) during extension not surpassing 1%, and (c) after further extension. ——, Tensile load-carrying bonds; load-carrying ability decreasing, – – –, and increasing, ···, with extension.

When the distance between atoms becomes overextended beyond the bond energy trough as shown in Figure 3.1 (p. 22), bond separation will occur and the specimen separates into two pieces. Since this break always occurs at the location of the weakest link, any flaws will let that break occur at a stress level considerably lower than calculated according to the known atomic bond strength values. This failure mode, described as brittle failure, is characteristic of nearly all ceramics and glasses, but also occurs with some metals and plastics.

Other mechanisms must exist for materials that will not fail under identical test conditions in a brittle mode. In the case of ductile metals the created void space can become replenished by a sideways motion of atoms, which would be equivalent to a slipping motion of rows of atoms. According to Figure 7.3 the broken bonds between atoms A and A', etc. would be reestablished between atoms B and A', etc. This change involves, in addition to the extension, a concomitant reduction in the specimen's width. If this process is repeated, an overall ductile deformation will have taken place. All the work will have been converted to heat, since the metal retains its high crystalline order and low entropy.

Because of their dominating strength in the direction of chain links versus the weaker intermolecular bonds, chain links will in plastics initially move in a longitudinal direction as the polymer chain becomes more aligned in the direction of pull (Figure 6.23, p. 172). Only a fraction of the work is converted to heat; most is consumed for orientation, causing large changes in entropy. The evenly strained specimen reaches a maximum stress (yield strength) where secondary processes allow further extensions to take place. A sliding action (shear yielding) between neighboring chains similar to the planes in metal crystals can take over on further stretching. In many plastics, necking will occur first in a confined region, then slowly extend over the entire specimen. The overall result will also be a reduction in cross section of the test specimen. At further extensions, when most of the easily orientable links have been rearranged, the tensile strength may increase again, particularly when the indicated force is divided by the now much smaller cross-sectional area. Ultimately the final separation of the test specimen will occur at the tensile strength-at-break level, which is by no means of importance for practical applications.

In other plastics, particularly the highly filled, crystalline or rubber modified materials, further extension is not accompanied by a sliding action but by crazing, an accumulation of narrow layers of semivoids arranged perpendicular to the direction of load (see also p. 194).

The character of the stress–strain curve greatly depends on how the atoms (in metals) or the chain links (in plastics) participate in the integrating sideways motion that takes place in a shear mode. Materials with high chemical bond strength and most highly cross-linked polymers (thermosets) exhibit a brittle behavior. The reader should be reminded that diamond with its extremely strong bonds extending in like manner into four directions is a very brittle material. The great ductility of polyethylene is based on the randomly transposed directions of the individual chain linkages both in regard to the polymer chain itself and in relationship to surrounding chains.

The initial portion of the stress–strain curve was dependent only on atomic structure, related to each material, and on temperature, which dominates atom mobility. The remaining part of that curve is governed by the great number of previously indicated influences.

Geometric constraints need to be emphasized again since they are so important for patterning any designed part. A material in the form of a thin fiber will have the best chance of providing toughness and folding endurance. In parts with heavy cross-sectional areas, the shear forces will attain such a magnitude that even the most pliable materials will fail by brittle fracture. Similarly, smooth surface and uniform cross-sectioned parts will promote ductile deformations, whereas rough surfaced parts and parts with notches or abruptly changing cross sections will tend toward a brittle failure.

Environmental influences (chemicals) will not affect the shape of the stress–strain curve at the outset, unless these substances become incorporated as plasticizers in the polymer. They may terminate the progression along the curve by promoting a separation of the strained specimen prior to the onset of yielding (shear motion), which results in an early, low-strength, brittle failure.

Hydrostatic pressure will often prevent brittle fracture by promoting plastic deformations that will preclude the materialization of tensile forces. In regard to the various forms of stress that favor a brittle versus a ductile break, the following tiers are listed in declining order: notched specimens tested under

FIGURE 7.4.   Polyethylene bottle ruptured in a brittle mode when filled with water and subjected to freezing temperature.

flexure, and specimens subjected to tensile forces, to shear forces, and finally to compressive forces.

It is quite perilous to generalize the failure mode of a certain material only by its overall behavior. Rubber, being a very flexible material, will fail by a brittle mode when under high stress. An air-inflated rubber balloon will split into several pieces when gently poked with a needle. Figure 7.4 shows how a polyethylene bottle, a few years old, can rupture in a brittle way when filled with water and subjected to below freezing temperature.

Practically any material can be brought to whatever desirable brittle or ductile form by the selection of appropriate circumstances, for example, forging of heated metals, explosive forming of some metals, cryogenic pulverizing of rubber, sharp bending of steam-treated wood, and blowing and forming of heated glass.

Still other important information can be taken from the progression of the stress–strain curves. Since the product of force times distance equals work or energy, the area between the curve and the baseline corresponds to the energy expended into the specimen during deformation. This is suggestive of the maximum energy a certain volume of material could be capable of absorbing in case of impact. Since this information ties in with the energy absorption during impact testing, quite reliable conclusions can be drawn as to the expected impact behavior of a particular material.

Under cyclic loading conditions, especially under reversed loading, the stress–strain curve will convert into a series of related mechanical hysteresis loops.

FRACTURE TOUGHNESS

The second most important material selection criterion is the establishment of the minimum toughness requirements. The difficulty in defining this value is illustrated by the question a designer who intends to replace a brass hinge with a plastic one must ask: How hard can a door be slammed before the hinges will separate? In this case, if a piece of wood jams the door at the right spot, even a child could separate the hinges easily.

The hinge example indicates that the determination of a material's toughness involves much more than the simple analysis of data on its strength. Whereas in a tensile test the material can be followed incrementally up to its point of failure, for a toughness determination a specimen of the material must first be rapidly destroyed with the event analyzed later.

IMPACT STRENGTH

Impact values represent the total ability of the material to absorb impact energy. This total is composed not only of the energy required to break primary and secondary bonds but also of the work consumed in deforming a certain volume of the material under test. Even with expensive equipment a clear separation of these two values is very difficult to establish.

Fracture toughness relationships have been adapted to plastics from fracture toughness tests developed for metals; these are described in the ASTM E 399 test procedure. However, they will apply only to a selective mode of failure (plane strain). Since the specimen must fail in a brittle mode, it must contain a very pointed notch and must be provided with a large cross section to prevent the material from separating in a ductile manner. These conditions rarely relate to actual plastics parts.

Many other mathematical formulas are cited in the literature dealing with fracture phenomena. Because of the complexity and multiplicity of events they are applicable only under narrow conditions and therefore are not yet suitable for general design purposes. Therefore, especially for large volume applications, service-related impact tests should be devised in nearly all cases.

Furthermore, at the outset, a determination must be made of the type of impact damage that is tolerable, intolerable, or desirable. This can be done in many ways. Establishing the probability of occurrence of destructive impact loadings, as is done in many applications where ordinary glass is used, would be one way. Establishing the probability of suffering permanent deformation and determining to what extent this is tolerable (e.g., slight denting of automobile frame members), would constitute another way. Establishing crashworthiness, a quite distinct impact property, with the aim of obtaining the highest energy absorption by sacrificing the part to protect more valuable components of an assemblage, would represent a third way.

Most impact tests are carried out by subjecting the test sample to a sudden bending force, since this relates best to actual impacting conditions for most structural parts. However, because of their simple mechanistic status, tensile impact tests give results that are much more suitable to mathematical manipulation.

The notched Izod pendulum impact test (ASTM D 256) is preferred for cases in which materials are selected for more compact parts, whereas the falling weight impact tests are more applicable for larger extended structures. In the falling weight impact test (ASTM D 3029) the material is placed under a multiaxial load, which is more akin to larger molded parts usually combining various flat shapes. Specimens tested according to the Izod impact test are subjected to only uniaxial forces. More variations in test results can be expected from variations in molding conditions, as specimens may exhibit greater or lesser molecular orientation or packing. The notched specimens incorporate very severe conditions that seldom exist in actual parts. However, because many of the thermoplastic engineering materials would not fail if tested as unnotched specimens, such testing would result in meaningless numerical values.

Only for specimens that fail in a brittle mode can it be concluded that the strength of the chemical bonds will be reflected in the observed energy values. In the case of ductile impact failure, the bulk of the absorbed impact energy will be used for the alignment of the chains. Therefore, very high impact strength values can be obtained only when a large volume of the material participates in that deformation process.

Depending on the rate at which these deformations occur, changes in brittle–ductile transitions may proceed in either direction depending on circumstances. The amount of energy absorbed in brittle specimens is nearly rate independent, whereas ductile breaking specimens will usually absorb less energy at very high rates of impact. In the case of toughened nylon a sudden transition to brittle failure takes place at a specific rate threshold. With rubber-modified acrylonitrile–butadiene–styrene plastics, which generally have a lower impact strength, toughness will diminish negligibly when rate of impact is raised.

Several theories exist relating to the start of an incipient crack and also to the propagation of growing cracks. The formation of a crack could remain limited to a single site or, as in many cases, lead to many crazes, of which only one will form the final crack.

CRAZING

Crazes can readily develop in the amorphous domains of plastics when subject to tensile forces. Although they are frequently observed in the presence of

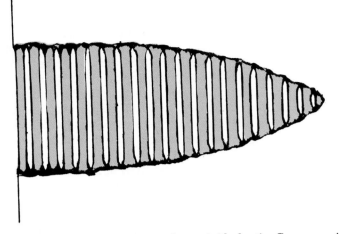

Figure 7.5.   Craze formation in an amorphous rigid plastic. Gray area indicates columns consisting of plastic material yielded by plastic flow.

liquids, they also develop in the presence of gases or vapors. Starting from the outside or around foreign inclusions, the elastically dilated matter will be converted into the craze substance consisting of highly drawn out fibrils, at a length of one to several hundred nanometers, evenly distributed between the voids, as seen in Figure 7.5. Although the craze matter consists of only about 50% plastic material, it is still capable of supporting an appreciable load.

The areas will quickly grow only in directions perpendicular to the direction of the tensile force. Additional crazes form in consecutive parallel rows at regular intervals, sometimes very closely together. Since there is a great difference in the index of refraction between the craze matter and the surrounding plastic, crazes can be easily identified by their shiny reflections of incident light (Fig. 7.6). Impact modified, pigmented, or filled plastics will usually turn white or brighter in color when crazes develop.

On further stressing, a separation of the specimen will occur through the crazed material, an observation verified by electron microscopic examinations of the brittle fracture surface. From Figure 7.7 it can be concluded that in the case of brittle fracture a very small amount of material will undergo ductile deformation. Although the amount of material subject to deformation is low, the formation of numerous crazes can increase energy absorption of brittle plastics over highly cross-linked thermoset resins by two orders of magnitude. Still, truly ductile breaks will be incomparably better energy absorbers.

Ductile–Brittle Transition

As explained in the discussion on tensile tests, the most profound differences in fracture responses occur when the failure mode changes from a brittle to a ductile mode. The arbitrary selection of specimen sizes has set many materials

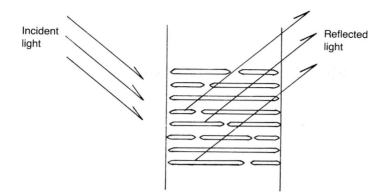

FIGURE 7.6.   Light reflection observable on crazed specimen.

in the region of consistently low-energy, brittle breaks and others in the region of consistently high-energy, ductile failures. A certain confusion has arisen for a few materials that, depending on other, sometimes minor variations, may fail in either a brittle or ductile mode.

This transition from a ductile to a brittle failure can take place under many different circumstances of test conditions, dimensions, or material variations. As has already been stated, a crack will nearly always start with a plastic deformation at a point or line of highest stress. Only very brittle materials such as the highly cross-linked thermosets and glasses will break in a truly brittle mode. In ductile materials yielding will continue throughout the whole cross section if the material is capable of abating the developing shear forces by narrowing toward the center of the piece. Figure 7.7 shows the cross sections through the centers of broken Izod impact bars illustrating the various degrees of ductility.

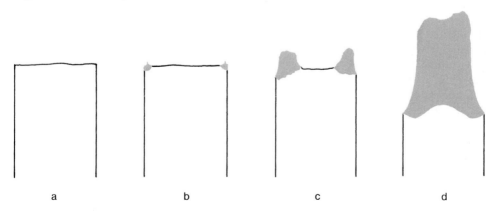

a   b   c   d

FIGURE 7.7.   Impact-tested specimens of various degrees of ductility. Longitudinal sections through the center of broken Izod impact bars with yielded material marked in gray. (a) Extremely brittle thermoset material; (b) brittle thermoplastic material; (c) partly ductile thermoplastic material; (d) high-impact thermoplastic material.

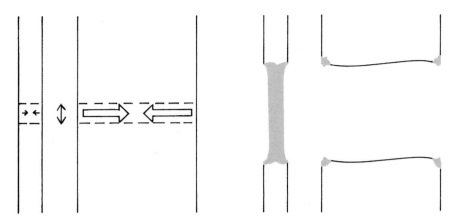

FIGURE 7.8. Ductile and brittle failure of the same thermoplastic material. Left: Sizes of arrows mark magnitude of tensile T and shear S forces to be expected at a hypothetical extension prior to failure in the case of most thermoplastics. Right: Only the thinner wall specimen will yield.

Experience has shown that the intermediate mode of failure as sketched in example (c) of Figure 7.7 occurs rather infrequently since once yielding has been established the process will continue over the full width of the sample.

Increasing specimen thickness will invariably lead in all materials to a brittle failure at one point as a result of the increasing shear forces as illustrated in Figure 7.8. This is also true for steel plates. Toughened plastics are much less susceptible to this effect since they can experience quite high elongations without lateral contraction because of the formation of crazes or voids. This is especially characteristic for acrylonitrile–butadiene–styrene copolymer plastics.

As indicated, the critical thickness at which ductile failure changes to brittle failure may vary from a few thousandths of an inch (about 0.1 mm) or up to an inch (several cm). In Figure 7.9 the test results obtained with polycarbonate specimens are plotted, giving a clear indication that the boundary for this material at room temperature is at about 0.170 inch (4.2 mm).

Curves exhibiting the same characteristics can be obtained by changing any other dimension in test specimens. Figure 7.10 demonstrates the sudden change from a ductile to brittle failure mode when only the *notch radius* of rigid polyvinyl chloride specimens is altered.

Similarly, by *changing the temperature* of the test specimens but keeping their thickness constant a corresponding response can be observed. Again using polycarbonate specimens of 1/8 in. (3.2 mm) thickness, their impact strength has been determined between $-65°F$ ($-54°C$) and $300°F$ ($149°C$). Under these conditions the boundary between brittle and ductile failure has been established at $8°F$ ($-13°C$) (Fig. 7.11).

Some of the other conditions that could cause ductile test specimens to fail in a brittle mode should be briefly mentioned. *Lower molecular weight resins* will

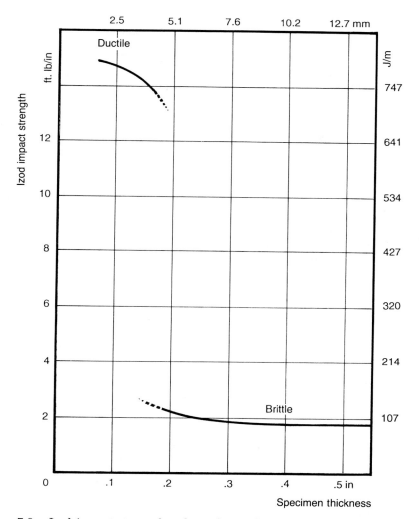

FIGURE 7.9.   Izod impact strength values obtained with polycarbonate specimens of varying thicknesses.

more likely turn brittle. The same results will be obtained whether one starts out with a lower molecular weight product or whether a breakdown in molecular weight occurs during molding (poorly dried material) or later during use. Unstabilized specimens that have been exposed to outdoor weather or to UV radiation might also become brittle because of breakdown of the polymer. Although in polycarbonate the ultraviolet rays will be completely absorbed within 0.03 in. (0.8 mm) from the surface, ordinary or heavy wall specimens of unstabilized resin will still fail in a brittle mode. However, a ductile break will again be experienced if that thin layer is removed by grinding or machining. This experiment also verifies that the mode of failure will be determined at

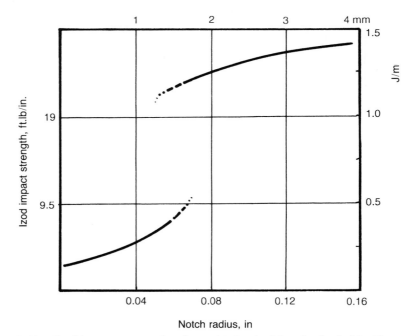

FIGURE 7.10.   Izod impact strength values obtained with polyvinyl chloride specimens with varying notch radii. Adapted from J.W. Summers and E.B. Robinovitch, Polyvinyl chloride. In Engineering plastics, Vol. 2. ASM International, Metals Park, 1988, p. 210.

the start of fracture. Once a specimen has started to fail in a brittle manner, the failure mode will stay locked in even if the central part consists of ductile material. This kind of behavior has regrettably been experienced when hard, brittle coatings were applied to ductile plastics. Conversely, soft coatings can sometimes provide enough protection to trigger a ductile failure mode.

On *annealing of specimens,* the metastable state of quickly cooled injection molded parts can become converted into a more stable, more compactly arranged molecular structure. This higher density material also tends to fail more in a brittle mode. That this phenomenon, customarily designated physical aging, does not represent an aging process is illustrated by polycarbonate. No embrittlement is found if parts are annealed at 300°F (150°C) for several weeks, although samples annealed at 255°F (125°C) will become brittle within a much shorter time. These brittle specimens regain ductility when "rejuvenated" at 300°F (150°C) for only a few minutes.

The impact energy absorbed in cases of ductile failure is always much higher than the values obtained with brittle specimens. Ductile and brittle failure values should therefore never be averaged because wrong information might be conveyed. For instance, the average impact strength value for a lot of safety helmets may satisfy the minimum requirements because of excellent impact strength values for half the specimens producing high ductile failures; however, the other half of the tested helmets may have failed in a brittle mode,

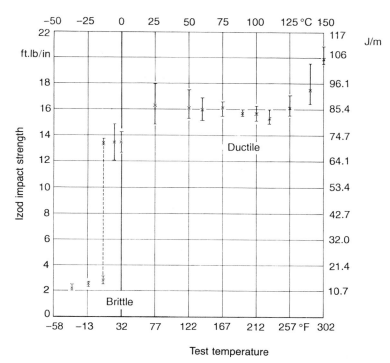

FIGURE 7.11. Izod impact strength values obtained with polycarbonate specimens when tested at different temperatures. From G. Gruenwald, Kunststoffe **50**, 383, 1960.

not providing any protection to the wearer. A statement of 50% low impact brittle failures gives the right message. The same caution about averaging also applies to the data expressed in Figures 7.9, 7.10, and 7.11, where in reality the two curves for brittle and ductile failures never meet.

Other circumstances influencing the impact behavior of thermoplastics are the degree of crystallinity, the size of the crystallites, and the chain orientation. Crystallinity will generally make the plastic more brittle because of its higher rigidity. However, its beneficial effects can be seen in a reduced tendency to form large crazes, especially with smaller size crystallites. Orientation is beneficial only when the direction of the impact force coincides with the direction of the orientation and if the part has not already been extended during the orientation process to its limit.

In reviewing the most important mechanical properties, stiffness and toughness, which can be attributed to any form of structural materials, it becomes clear that it is very difficult to achieve top performance in both areas. As shown in Figure 7.12, corresponding values of the most common plastic bulk materials will fall within the crescent-shaped area indicated. Materials positioned within the desirable area, marked by a dashed line, are still missing. To cover that area it is necessary to incorporate specifically laid out structural reinforcements such as applied-in-advance fiber-reinforced composites or the well-ordered depositions of cellulose molecules in wood. In the case of metals,

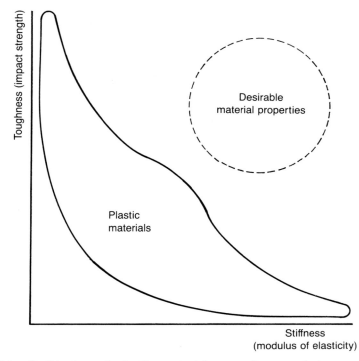

FIGURE 7.12. Positioning of plastic materials according to their toughness and stiffness.

careful selection of microstructure, hardness, and other attributes can approach that state. Transfigured to large size structures—such as the Eiffel Tower—those overall properties can be obtained by judicious positioning of the structural materials in struts and trusses.

Still one must be on the alert for deficiencies in other important material requirements. Everyone who has tried to split a tree trunk where either big branches or several roots interweave can attest to the high strength and toughness of those parts of a tree. Still, those pieces cannot be used for furniture since they will warp and twist, depending on changing humidity conditions.

There are other material properties that must be listed under mechanical properties and that are of great importance in the selection of materials for sometimes very specific applications. Their significance is described in various parts of this book. Some of them, such as hardness, coefficient of friction, and surface wear, are more closely related to surface properties and their performance may, in many cases, be enhanced either by surface treatments, coatings, or special additives.

Many readers will undoubtedly miss a tabulation of materials properties in this book. It has always been very expedient to look up property tables to obtain a quick property profile to assist in comparing basic plastic materials. This has been true as long as just a few basic plastics materials fulfilled the needs for most plastics applications. At present, and probably even more so in the future, basic plastics are modified by copolymerization, impact modification, and blending (not even taking into account the many reinforcements) to such a great extent that the originally characteristic material properties do not prevail anymore. Cited should be the listing of Izod impact strength values in handbooks for rigid polyvinyl chloride extending from 0.2 to 22 ft lb/in. (10 to 1200 J/m). This development is not limited to mechanical properties but extends into other fields. Many traditional opaque resins are now also available in transparent and translucent versions. This makes it imperative not to rely on generic, encompassing data but to peruse the properties sheets provided by the manufacturer for the material that is intended to be used, while still considering the many admonitions cited in this chapter.

References

M.F. Ashby and D.R.H. Jones, Engineering Materials. Pergamon Press, Oxford, 1980.

W. Brostow, Science of Materials. John Wiley, New York, 1979.

W. Brostow and R.G. Corneliessen, Failure of Plastics. Hanser, Munich, 1986.

C.B. Bucknall. Toughened Plastics. Applied Science, London, 1977.

W.D. Callister, Jr., Materials Science and Engineering. John Wiley, New York, 1985.

J.D. Ferry, Viscoelastic Properties of Polymers, 2nd ed. John Wiley, New York, 1970.

J.E. Gordon, The New Science of Strong Materials, 2nd ed. Princeton University Press, Princeton, NJ, 1976.

J.E. Gordon, Structures, or Why Things Don't Fall Down. Plenum Press, New York, 1978.

C. Hall, Polymer Materials, an Introduction for Technologists and Scientists, 2nd ed. John Wiley, New York, 1989.

R.P. Kambour, Crazing. In Polymers: An Encyclopedic Sourcebook of Engineering Properties. J.I. Kroschwitz, ed., p. 152. John Wiley, New York, 1987.

H.H. Kausch, Polymer Fracture. Springer-Verlag, Berlin, 1987.

H.H. Kausch and G. Williams, Fracture. In Polymers: An Encyclopedic Sourcebook of Engineering Properties. J.I. Kroschwitz, ed., p. 365. John Wiley, New York, 1987.

I.M. Ward, Mechanical Properties of Solid Polymers, 2nd ed. John Wiley, Chichester, 1983.

I.M. Ward, Mechanical Properties. In Polymers: An Encyclopedic Sourcebook of Engineering Properties. J.I. Kroschwitz, ed., p. 569. John Wiley, New York, 1987.

A.F. Yee, Impact resistance. In Polymers: An Encyclopedic Sourcebook of Engineering Properties. J.I. Kroschwitz, ed., p. 476. John Wiley, New York, 1987.

8

Thermal Properties of Polymeric Materials

No other group of materials is as affected by temperature changes as are plastics. Unquestionably, any structural material will, at an elevated temperature, lose its ability to support a load and most will, at a still higher temperature, convert to a more or less viscous melt. But these temperatures are extremely high if compared with many plastics materials. Only a very few of the metallic or ceramic-type materials will show changes in properties between $-40°F$ ($40°C$) and $200°F$ ($100°C$) (e.g., brittle fracture of some steels at the low-temperature range). In the case of nearly all plastics, their properties at these arbitrarily selected temperature margins become noncomparable. What is even more significant is that at around room temperature all metals (except lead) and ceramics are so stable that creep and stress relaxation do not have to be considered. A general rule states that creep, a time-dependent deformation of loaded structures, will occur only at temperatures at least above half the material's melting temperature in degrees Rankine (Kelvin). Therefore problems such as viscoelastic behavior must always be considered by all plastics practitioners. Only a few metallurgists working on high-temperature turbine and chemical process apparatus components have to share the same concerns.

MELT SOLIDIFICATION

To obtain an understanding of the changes taking place in materials at various temperature levels, our examination begins with the state of a melt. Since the atoms in molten metals exhibit free mobility, they are free to rotate, vibrate, and translate, and since they consist of small entities, their viscosity is very low. The same is the case for the cations and anions in salts or the smaller molecules in most organic compounds. Some salts and most ceramic-type materials will be present in the melt as voluminous ionic aggregates or ion clusters, which will moderate the mobility of these particles. These melts might therefore also have very high viscosities. Although the chain links of thermoplastic polymeric materials possess free vibration, their rotation and translation will be curtailed because they will remain connected on two sides to other links of the macromolecule. This too will give the melt a high viscosity, which will mainly be dictated by the length of the polymer chain (e.g., the ratio of number of chain links to number of chain ends).

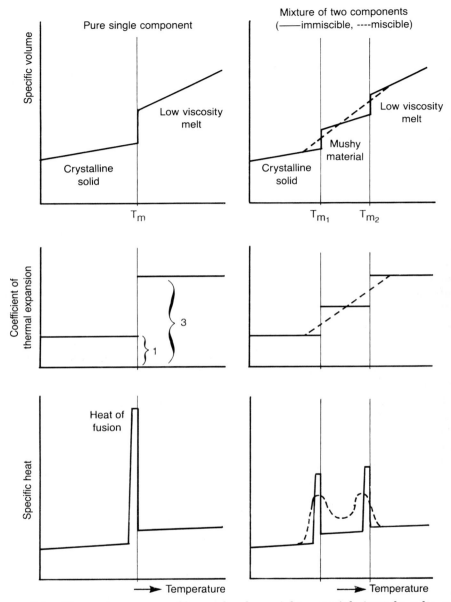

FIGURE 8.1.   Thermal responses of low-molecular-weight materials (metals, salts, and organic compounds).

In the melt all thermal relationships are quite straightforward. As can be seen in the following tables, showing the decreases in specific volume (the reciprocal of specific gravity) with decreasing temperatures, a straight line characterizes the rate at which thermal contraction takes place (coefficient of cubic thermal expansion) and also the rate at which heat is simultaneously dissipated (specific heat). Figure 8.1 shows schematically the curves—which must be read from right to left—for low-molecular-weight organic compounds, metals, and salts. The specific volumes of all melts will recede while dissipating heat and lowering the temperature of the molten material. This holds true for the pure compounds as well as for a soluble mixture of metals or a mixture of organic compounds.

The changes in melt viscosity with melt temperature will be minor in comparison to the massive changes occurring during solidification. When these materials reach melting temperatures, designated T_m and also called first-order transition, the temperature will remain constant for as long as all material has solidified. The large amount of heat dissipated at that temperature corresponds to the heat of fusion. The solid represents a stable crystalline material that, on further cooling, loses less in specific volume at even lower specific heat values. On the right-hand side of Figure 8.1 the same curves are repeated for mixtures of salts or organic compounds or metal alloys. Depending on how miscible, interreactive, or soluble both components are, an intermediary mushy state may extend beyond the two melting temperatures and the respective curves may mutate from a step-like line to a more rounded shape.

The thermal responses of amorphous yet polymeric materials, whether a silicate glass or an amorphous thermoplastic resin, are illustrated on the left-hand side in Figure 8.2 subject to the same circumstances. These materials have high melt viscosities as a result of the encircling amassments or the entangling long chains. The purely amorphous materials exhibit an uncharacteristic change when they cross the temperature region where their melting temperature would be expected to lie. Below this region the viscosity is so high that melt processing of these materials becomes impossible. During cooling their viscosity builds up when neighboring chains increasingly approach each other. The generated bonds do not become locked into a crystal lattice but may first provide a temporary connection switching back and forth to other segments. Since these materials are not capable of crystallizing, but also are not free to flow anymore, they are considered to be in a state of a cooled, viscoelastic melt, which exhibits an elastomeric-like behavior, as long as the temperature stays within the rubbery plateau. As the temperature is lowered further, the bonds become more lasting. Within this span some segments (groups of atoms) may retain much of their ability to vibrate and, to a limited extent, to rotate and translate, whereas others may become restricted just to vibrational motions. Starting from that point the specific volume curve deviates from the straight line of the melt and on further cooling the ensuing course of the track becomes very much time dependent. This means that at different cooling rates the shapes of the curves in that region will change.

Because of the very high viscosity, solidification starts before the groups of long chain or ring segments can be arranged in such a way that they occupy minimum space. Therefore—at practical cooling rates—a small amount of volume, amounting to about 2%, will remain unoccupied. This "free volume" may affect some properties. The argument whether this represents the cause or the effect of the discrepancy in specific gravity is immaterial. However, the equilibrium condition representing the most stable state (lowest specific vol-

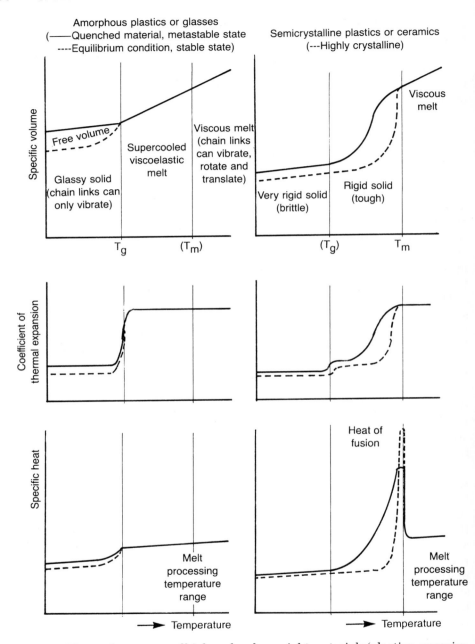

FIGURE 8.2.   Thermal responses of high-molecular-weight materials (plastics, ceramics, and glasses).

ume) can only partly be obtained by a physical aging process or some other methods as described in Chapter 9. The solid line curves in Figure 8.2 represent the specific volume values obtained under practical cooling rates, whereas the dashed line curves could be obtained only under extremely low cooling rates, mainly in the vicinity of the glass transition temperature. The gap between both lines can be narrowed by annealing—extended exposure time at an elevated temperature.

GLASS TRANSITION TEMPERATURE

By extending the two straight lines of the two specific volume curves to the crossover point, the *glass transition temperature,* sometimes still called the second-order transition temperature, T_g, is obtained. This point can also be derived from other measurements (e.g., the midpoint of the step change from the two characteristic specific heat values or from optical measurements such as index of refraction or from viscosity-related determinations). Internal segmental motions in polymers are also studied by dynamic mechanical and nuclear magnetic resonance techniques. A very high viscosity of 10^{11} to 10^{13} P (10^{10} to 10^{12} Pa·s) is generally regarded as a reasonable region to separate liquid from solid behavior, which represents the characteristic attribute for the glass transition temperature. It is important to realize that the glass transition temperature is neither a thermodynamic transition such as the melting transition nor a sharp temperature point.

Although the glass transition temperature is regarded as a characteristic material constant, unfortunately, depending on the test conditions, quite divergent values, which can be spread out over 50°F (30 K) are obtained. At slow cooling rates the glass transition temperature materializes at markedly lower temperatures. The degree of crystallinity and the molecular weight dispersity will also have an effect. At the low-molecular-weight end it will drop sharply.

An additional complication arises when semicrystalline plastics are investigated, especially when their rate of crystallization is slow. The corresponding curves for materials of that kind are sketched on the right-hand side of Figure 8.2.

It should be quite plausible to assume that heterogeneous polymers, such as nylons, which consist of a kind of polyethylene with regularly interspersed amide groups, will have two regions where neighboring molecules can relate: one at the amide group where hydrogen bonds are formed and another where the more abundant methylene links become attracted by the much weaker van der Waals forces (see Fig. 6.19 on p. 164). Although they have only one glass transition temperature because of the above-stated definitions, some minor irregularities may be seen in the specific volume–temperature chart. For a substance to exhibit two glass transition regions, two separate phases such as those found in some blends and alloys must be present. But this discrepancy will not remain inconsequential as will be shown later. The great differences in the deflection temperatures under the two different loadings listed in Table 8.1 prove that. But even very simple and symmetrical polymers such as poly-

TABLE 8.1.   Thermomechanical Data of Various Plastics

Plastic	Deflection Temperature under Flexural Load				Glass Transition Temperature		Crystalline Melting Temperature	
	At 264 psi (1.82 MPa)		At 66 psi (0.45 MPa)					
	°F	°C	°F	°C	°F	°C	°F	°C
Polyethylene, high density	—	—	185	85			275	135
Polypropylene	130	55	240	115	−4	−20	335	170
Polymethylpentene	120	50	185	85			450	232
Ionomer—ethylene copolymer	95	35	120	48			190	88
Polystyrene	210	100	210	100	220	105		
Polyvinyl chloride	160	70	160	70	210	100		
Polymethylmethacrylate	205	95	215	100	210	100		
ABS, heat resistant	230	110	240	115	250	120		
Acetal, homopolymer	260	125	330	165	−120	−85	350	175
Acetal, copolymer	240	115	320	160			330	165
Nylon 6	165	75	370	188			420	215
Nylon 12	115	45	210	100	280	140	360	180
Nylon 6/6	175	80	450	230			500	260
Cellulose acetate butyrate	160	70	180	80			280	140
Polyethylene terephthalate	85	30	300	150	170	75	490	255
Polybutylene terephthalate	150	65	270	130			470	240
Polyphenylene oxide, modified	220	105	270	130	240	115		
Polycarbonate	270	130	280	135	300	150		
Polyphenylene sulfide	275	135	—	—	190	88	550	288
Polysulfone	345	175	360	180	375	190		
Polyethersulfone	395	200			440	225		
Polyetherimide	390	198	410	210	420	215		

tetrafluoroethylene can present a series of transition stages. These transitions are in the case of polytetrafluoroethylene most likely based on its many different crystal structures (comparable to the multitude of metallic crystal structures recognizable at specimens crystallized at different cooling rates and temperatures). The inclination of the straight line in the melt region also defines the coefficient of cubic expansion, as the slope of the straight line at the other side (solid) refers to the coefficient of linear expansion, appearing at a 3:1 ratio.

The remarkable difference in the shape of the overall curves in Figures 8.1 and 8.2 gives a good indication that an applicable description about the mechanical behavior of amorphous and semicrystalline polymers could not be reached based only on glass transition temperature values. Table 8.1 lists thermomechanical data characterizing various groups of plastics.

HEAT DEFLECTION TEMPERATURE

For practical applications the glass transition temperature represents the upper temperature limit for amorphous structural plastics. Their *heat deflection temperature* at the high load is generally 15–35°F (10–20 K) below the glass transition temperature. For elastomers the glass transition temperature represents the lower temperature limit for their application. The highly semicrystalline plastics, on the other hand, are applicable close to their crystalline melting points. Because of that ambiguity and the need for plastics engineers to know the boundary temperatures for any mechanical application of plastics, tests have been devised that will provide values for these practical temperature constraints. The load-carrying capability was first arbitrarily selected at a rather high value of 264 psi (1.82 MPa), which is just about one-quarter of the limit to which unreinforced structural plastics can safely be loaded. When it was noticed that nylon resins produced ridiculously low temperature limits, the loading was again cut to one-quarter, 66 psi (455 kPa), whereby the temperature values for nylon resins climbed to the top.

One must always remember that the listed heat deflection temperature just marks one point of a curve, the point at which the 4-in.(10.2-cm)-long test bar deflects 0.01 in. (0.25 mm). In Figure 8.3 experimentally derived curves for a typical semicrystalline and an amorphous resin are drawn.

A further modification is sometimes mentioned when ranking the heat deflection behavior of plastics. When the temperature of molded parts is slowly increased in an oven, at a certain temperature threshold, the parts slump because of the rather small gravitational force. This zero load heat deflection temperature will again position the plastics in a different order. This temperature is closest to the crystalline melting temperature and therefore benefits semicrystalline plastic. Compare the data given in Figure 6.17 (p. 161).

The rating of the thermal capability of plastics materials when used in electrical equipment has been resolved by the Underwriters' Laboratories in document UL746C. Under these conditions all the mechanical, electrical, and thermal properties have been taken into consideration. In the following tab-

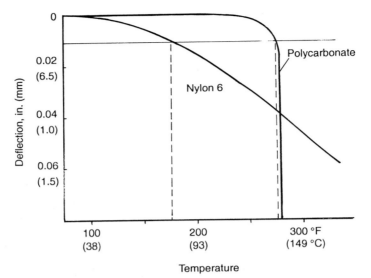

FIGURE 8.3.   Curves showing the deflection of 4-in. (10.2-cm) span bars, 0.5 in. high (12.7 mm), when heated 3.6°F/min (2 K/min) under a 264 psi (1.82 MPa) flexural load.

ulation the use temperatures—also called thermal index values—of many important plastics are listed. For comparison, temperature limits for some metals widely used in structural parts are also cited. These temperature limits are valid for applications extending over a long period of time and should not be confused with the short time thermal property expressed by the heat deflection temperature under load.

Upper Use Temperature		
°F	°C	Structural Materials
140	60	Acrylonitrile–butadiene–styrene
149	65	Polyamide, polypropylene, molded polyethyleneterephthalate parts
167	75	Polycarbonate, polybutyleneterephthalate
194	90	Epoxies, vulcanized fiber, wood, paper
212	100	Ureas
		Die cast zinc #3 and #7
221	105	Silicone rubber, polyethyleneterephthalate film, sheet molding compounds
266	130	Melamine resins, unsaturated polyester resins
302	150	Polychlorotrifluoroethylene, fluorinated ethylene propylene copolymer, polysulfone, phenolic resin, silicone molding compound
356	180	Polytetrafluoroethylene, polyethersulfone
401	205	**Aluminum A 380**
464	240	Polyetheretherketone
1200	650	**Nonalloyed carbon steel**

As was stated in Chapter 7, most mechanical properties are determined only at room temperature, which could be 75°F (24°C) to 68°F (20°C). These posted properties values will change with increasing temperature until the listed boundary temperatures of no strength are reached. The slope can be quite different and can fluctuate. The presence of fibrous reinforcements raises the mechanical properties and diminishes the drop-off observed with increasing temperature.

The incorporation of 40% glass fibers produces the following increases in the deflection temperature under a 264 psi (1.82 MPa) load:

Plastic	Heat Deflection Temperature of Neat Resin	Increase over Unfilled Resin
Polycarbonate	295°F (146°C)	25°F (15 K)
Polybutyleneterephthalate	400°F (205°C)	250°F (140 K)
Nylon 6/6	480°F (250°C)	305°F (170 K)

Again the great differences between amorphous and semicrystalline polymers become apparent.

SECOND-ORDER TRANSITION

When temperatures are lowered, the modulus of elasticity of all plastics will increase. Since elongation values will decrease at a higher rate, all plastics and even rubbers will eventually become hard and very brittle. At the *second-order transition point*, T_β, often described as the beta-transition temperature, the mobility of all the chain links will become limited to minor vibrations (no rotation or translation possible) making any yielding of chain segments impossible. Below that temperature many plastic materials will have only limited serviceability. Since these temperatures are quite low and since many plastics have—as already indicated—several transition points and are affected at variable degrees, these values are not well publicized. The ductile–brittle transition temperature (see p. 196 (text)/199 (graph)) is much higher than the beta-transition temperature, which should more properly be called the brittle–superbrittle transition temperature. Just two impressive examples should be given here since, in other instances, for materials considered to be quite brittle (polystyrene and polymethyl methacrylate), the second-order transition temperatures lie above room temperature.

Polymer	Glass Transition Temperature, T_g		Second-Order Transition Temperature, T_β	
	°F	°C	°F	°C
Polyvinyl chloride	175	80	−58	−50
Polycarbonate	300	150	−185	−120

To now only some general properties, mainly mechanical, that are greatly affected by temperature have been described. The truly thermal properties that are by themselves characteristic for each material should be regarded next.

Thermal Conductivity

The *thermal conductivity* of plastics is important not only for managing the various plastics processing steps efficiently but also for rendering the thermal conductivity properties of many plastics appropriate for various demanding applications.

In general, as is the case for electrical conductivity, when compared with other structural materials plastics are situated at the low end in regard to thermal conductivity. Again metals are at the extreme high side and mineral or ceramic-type materials are mostly in between. Since heat is conducted prevailingly along strong bonds, the structures of diamond and cristobalite (p. 12) explain why diamond represents the best thermal conductor, although it is an electrical insulator, whereas some minerals and ceramics with wide voids in their structure are good thermal (and electrical) insulators. The best metal, silver, has a thermal conductivity of 240 Btu/ft·hr·°F (420 W/m·K); because of its compactness and crystalline regularity, a pure carbon 12 diamond has an even higher thermal conductivity, an astounding 3000 Btu/ft·hr·°F (5000 W/m·K). Some characteristic thermal data for various materials are compiled in Table 8.2.

For applications in which extremely low thermal conductivities (good thermal insulation) are required, plastics can easily be modified by converting them to foams. The thermal conductivity of such foams may be as low as 0.01 Btu/ft·hr·°F (0.02 W/m·K). Depending on the volume fraction of plastic to air or vapor and on the geometry and structure of the voids, a required balance between mechanical strength and insulation properties can be struck. In other applications in which dissipation of heat (e.g., frictional heat in plastics bearings) through a thick layer of plastic would be insufficient, heat conductivity can be increased by the incorporation of mineral fillers or reinforcing fibers or even more by metal powders, fibers, and flakes. However, only with very high loadings will thermal conductivities approaching those of metals be obtained.

The thermal conductivity values for the neat plastics polymers are all within a narrow range of 0.12 to 0.2 Btu/ft·hr·°F (0.2 to 0.35 W/m·K) with just a few exceptions for high-density polyethylene, which at the upper end exhibits 0.3 Btu/ft·hr·°F (0.5 W/m·K). Polypropylene and polystyrene mark the lower end of the range at 0.07 Btu/ft·hr·°F (0.12 W/m·K).

Thermal Diffusivity

Whereas thermal conductivity describes the behavior of materials when exposed to different, but constant temperatures on each side, a variant of it, *thermal diffusivity,* relates to all thermal responses occurring in materials when temperature changes are involved. Practically, thermal diffusivity is inversely proportional to the time requirements for both

1. heating all plastics materials during the first part of plastic processing (melting) and
2. cooling all thermoplastics materials in the mold after forming.

By combining three important properties—thermal conductivity, density, and specific heat—thermal diffusivity can be used to rate thermoplastics in regard to the speed with which heat in the plastic can be added (by means of a heated cylinder) or extracted by a cooled mold. However, diffusivity values are seldom listed in properties sheets since their magnitude is greatly dependent on the range of the temperature span selected. All three material "constants" contained therein do not remain constant over an extended temperature range of the plastic.

Though melting or heating of plastics is not restricted to supplying heat externally, heating of plastics may also be accomplished internally by generating frictional heat (by extruder screws) or by dielectric heating (high-frequency heating of thermoset molding compounds). Cooling, however, is restricted to the outer surfaces of formed plastic parts and is therefore solely dependent on thermal diffusivity.

SPECIFIC HEAT

The *specific heat* values of all plastics lie between those of the metals and that of water, which is one, in both the British and the old metric system. Again, the spread among the various neat polymers is rather narrow and ranges from 0.3 to 0.6 Btu/lb·°F (1.2 to 2.5 kJ/kg·K). Compounded plastics may exhibit greater variations influenced by the contributions of the compounding ingredients.

HEAT OF FUSION

The low-energy content of crystalline solids causes the absorption of an appreciable amount of heat by the system during the melting process. This heat is called *heat of fusion* and varies only between 50 and 100 Btu/lb (125 and 250 J/g) for most thermoplastics. For comparison, the values for ice are 144 Btu/lb (335 J/g). The weight fraction of crystalline material in the plastic is of greater consequence for the amount of heat absorbed than the chemical composition of the plastic used. An equal amount of heat must be extracted during solidification, retarding the molding cycle at both the heating and cooling step.

LINEAR COEFFICIENT OF THERMAL EXPANSION

The *linear coefficient of thermal expansion* also represents a thermal property that is of significance in more than just one way. To obtain the correct dimen-

TABLE 8.2.   Thermal Properties for Various Materials Compared with Plastics

| | International system of units (SI) values | | | | English system values | | | |
| | Thermal Conductivity | | Specific Heat | | Coefficient of Linear Thermal Expansion | | Heat of Fusion | |
	W/m·K	Btu/ft·hr·°F	kJ/kg·K	Btu/lb·°F	m/m·10^{-4}/K	in/in·10^{-4}/°F	kJ/kg	Btu/lb
Copper	390	225	0.38	0.09	0.18	0.10	205	88
Aluminum	230	135	0.9	0.21	0.25	0.14	395	170
Stainless steel	15	10	0.5	0.12	0.15	0.10	—	—
Iron	70	42	0.46	0.11	0.11	0.06	265	115
Hardwood								
Parallel	0.15	0.08	2.2	0.5	0.05	0.03	—	—
Perpendicular	0.08	0.05	2.2	0.5	0.5	0.3	—	—
Cellulose acetate	0.25	0.15	1.6	0.4	0.8	0.8	—	—
Polyethylene								
High density	0.45	0.25	1.9	0.45	1.0	0.6	250	100
Low density	0.35	0.20	2.3	0.55	1.6	0.9	250	100
Polystyrene	0.12	0.07	1.2	0.3	0.7	0.35	—	—
Polycarbonate	0.18	0.1	1.25	0.3	0.7	0.35	—	—
Nylon	0.2	0.15	1.6	0.4	0.9	0.5	58	25
Phenolics, filled	0.5	0.2	1.2	0.3	0.3	0.2	—	—
Plastic foams	0.02	0.01	1.2	0.3	1	0.7	—	—
Air	0.025	0.015	1.00	0.24	—	—	—	—
Water	0.6	0.35	4.19	1.00	0.7	0.4	—	—
Ice	2.1	1.25	2.1	0.5	0.5	0.28	335	144
Diamond	Up to 5000	Up to 2900	0.52	0.125	0.012	0.007	—	—

TABLE 8.2. (Continued)

	International system of units (SI) values				English system values			
	Thermal Conductivity		Specific Heat		Coefficient of Linear Thermal Expansion		Heat of Fusion	
	W/m · K	Btu/ft · hr · °F	kJ/kg · K	Btu/lb · °F	m/m · 10^{-4}/K	in/in · 10^{-4}/°F	kJ/kg	Btu/lb
Silicones	0.2	0.1	1.5	0.36	5	3	—	—
Quartz glass	1.4	0.8	0.75	0.18	0.005	0.003	237	102
Glass	0.8	0.5	0.8	0.2	0.09	0.05	—	—
Porcelain	1.5	0.9	0.9	0.2	0.06	0.03	—	—
Concrete	0.5	0.3	0.8	0.2	—	—	—	—

sions of precision molded parts, mold shrinkage must be taken into account. Its main component is caused by thermal shrinkage, a material property of the solidifying plastic, and by the crystallization process of semicrystalline plastics. Another integral portion is related to molding conditions and to the layout of the mold and the gates. Parts solidifying over large cores will generally display lower shrinkages. The dimensions of mold cavities must be adjusted correspondingly case by case.

Other difficulties can arise when two different materials are used together in one assembly or when metallic parts are used as molding inserts. If there is an appreciable difference in their linear coefficient of thermal expansion either gaps or high forces (proportional to the modulus of elasticity of the weaker material) may develop when subjected to thermal cycles. In most cases damage to the plastic part can be averted by incorporating wider tolerances and reducing tightness of assembly fasteners (providing extension slots and applying a lower torque to threaded connections). Although similar changes in dimensions may occur in some kinds of plastics because of water absorption, one must consider that changes in temperature will become manifest within a few minutes whereas exposures extending over days or weeks are required for humidity-caused dimensional changes to take place.

The spread in linear coefficients of thermal expansion values is much larger in plastics than in metals, which range from 0.07 to 0.13×10^{-4} in/in·°F (0.12–0.24×10^{-4} m/m·K). In general most plastics fall within the range of 0.3–0.6×10^{-4} in/in·°F (0.5–1.0×10^{-4} m/m·K). Much lower values are obtained with high-temperature, highly aromatic polymers and highly filled fiber-reinforced composites. The polyethylenes, the cellulosics, and ionomers have about twice the expansion—0.6–1.0×10^{-4} in/in·°F (1.0–1.7×10^{-4} m/m·K)—of the majority of plastics. The silicones are in a group of their own with 1.5–4.5×10^{-4} in/in·°F (3.0–8.0×10^{-4} m/m·K), contrasting very much with the low values for fused silica, 0.03×10^{-4} in/in·°F (0.05×10^{-4} m/m·K), and glass, 0.5×10^{-4} in/in·°F (0.9×10^{-4} m/m·K), respectively.

The low values for fiber-reinforced plastics are based on the significant differences in the values of the coefficient for the matrix resins and the reinforcing fibers. Therefore thermal fatigue (shear forces acting at the fiber and matrix surfaces) must always be considered as a failure mode for composites that are regularly exposed to wide temperature fluctuations. This type of failure manifests itself by the formation of microcracks. There will also be a disparity in the coefficient of thermal expansion values depending on the direction of the fiber reinforcement. The dissimilarity in regard to direction also becomes manifested in wood. Its coefficient of thermal expansion is about 10 times higher in the transverse direction than along the grain, 0.15–0.3×10^{-4} in/in·°F (0.3–0.6×10^{-4} m/m·K) versus 0.015–0.03×10^{-4} in/in·°F (0.03–0.06×10^{-4} m/m·K). A certain part of this disparity is also the result of the lower coefficient of thermal expansion along covalent bonds, in contrast to the much weaker van der Waals bonds. Therefore, thermoset plastics with prevailing covalent bonds extending in all directions will have a lower coefficient of thermal expansion than thermoplastic resins. Similarly, highly oriented thermoplastic parts will extend less in the direction of orientation than in the transverse directions.

THERMAL STABILITY

Thermal stability is of great importance for all melt processable plastics. Chemical reactions proceed much faster in a material in a liquid than in a solid state. The gap between melt processing temperatures and thermal decomposition temperatures for high-temperature resistant plastics might become quite narrow. In a few instances (polytetrafluoroethylene and polyimides) the melting temperature is higher than the decomposition temperature, excluding any of the melt processing methods for their fabrication. Overheated plastic melts can decompose spontaneously, but in most cases impurities in them or chemicals (oxygen, water vapor) in the environment will accelerate decomposition markedly.

The decomposition products obtained will depend on whether the polymer will break somewhere along the chain, reducing average molecular weight rapidly, or split off monomer units from the chain ends, recognizable by the emanation of the monomer's characteristic odor. This so-called unzipping of the polymer can be suppressed by a judicious selection of stable chain end groups.

The general rule that the rate of a chemical reaction will double when the temperature is increased by about 18°F (10 K) is also applicable to plastics. This allows—with some caution—testing of parts for thermal degradation by exposing them to a higher temperature but for a more practical shorter period of time. The failure point is usually recognized when the tested property has reached about 50% of its original value. For the determination of the maximum service temperature or the long-term resistance of plastics the exposure time is usually selected at 10,000 hours. Similar criteria are employed for the evaluation of electrical insulation materials.

FLAMMABILITY

Ceramic-type materials that consist only of oxides have no problems in regard to *flammability,* and many of them are used for furnace linings. All other types of structural materials are capable of slow or rapid heat-generating oxidation reactions. Metals (especially magnesium and aluminum) in fine dispersion have caused fires as well as explosions, but more compact metal parts are quite safe since they will melt and thus be removed from high-temperature spots before combustion can proceed.

Since most plastics consist of organic chemical materials they are subject to violent oxidation reactions in the presence of air at high ambient temperatures, just like all common organic materials (e.g., wood, paper, textiles, and dry food products). Wood and other cellulose products (paper and fibers) represent solid materials that are most easy to ignite. They occur not only in a disaggregate but also in a nonfusible form. Commercially available plastics can span a wide range in regard to ease of ignition. Other aspects such as heat of combustion or char formation will also affect overall flammability of plastics. Because of many preconceived notions, the likelihood of plastics becoming involved in fires seems to be higher. Objects made out of plastics have generally much smaller wall thickness than wooden parts, and plastics prevail in thin sheet and film applications.

The weight fraction of hydrogen atoms in polymers is the greatest contributor to flammability, followed by the carbon content. For ignition to occur, organic materials must first be thermally degraded to flammable vapors. Halogen atoms (chlorine and bromine) are efficient flame retarders because of their high atomic weights and their ability to hinder the chemical combustion process. Sulfur may be a contributor to the combustion process in vulcanized rubber and the ethylene polysulfide-type elastomers. In the case of the phenylene sulfide polymers, this plastic's high melting and combustion temperatures based on the high content on benzene rings are providing some protection. In the polysulfone plastics the sulfur is contained in an already oxidized form (SO_2) promoting superior flame resistance. Similarly, a portion of the carbon atoms in polyester and polycarbonate polymers consists of already oxidized carbon atoms (CO_2). Advantage is also being taken of other beneficial ingredients and nonflammable fillers and reinforcing fibers to reduce flammability.

Fibrous reinforcements can sometimes become deleterious when they act like wicks, increasing the surface-to-volume ratio significantly (e.g., paraffin candles). Also overhead glass fiber-reinforced plastic ceiling panels are—in case of fire—prevented from melting and falling to the floor and thus remain exposed to the advancing flames, whereas unreinforced thermoplastic panels soften and drop to the floor, where they are less exposed to hot flames.

Most plastics are now also available compounded with flame-retardant and smoke-suppressant ingredients. Since these contribute to higher costs, impair some properties, and make processing more difficult, they are used only to the extent necessary.

Although several flammability tests have been standardized over the years, the search for actual fire-related tests is still going on. In most cases favorable flammability data are still given with a caution not to regard them as a guarantee for performance under actual fire conditions. This caution is also extended to the oxygen index values, even though they are very well suited for grading materials. The widespread belief that a material with an index above 20% (the oxygen content in air) will not propagate fire does not hold true. Under high temperature conditions, materials having significantly higher values will still readily burn.

For fire risk assessments there are at least five different conditions that can have decisive effects:

1. the ease of ignition (furniture upholstery, curtains, etc.);
2. the spatial distribution and part geometry controlling flame spread (bundles of electric wires, large thin-walled parts);
3. the total heat release, which is related to the amount of combustible materials expressed in pounds per cubic feet of inhabited space;
4. smoke obscuration, which might block the location of appropriate escape routes; and
5. the potential toxic hazards of combustion products.

In many cases legislative rulings dictate which of the various government specified tests must be passed. Submitting samples for testing to an independent testing laboratory is essential. In cases in which specific rules are not provided, one should still be cognizant of potential hazards surrounding the application of the molded part and at least communicate with the materials supplier to ensure that the most suitable formulation for any particular application is secured.

REFERENCES

W.D. Callister, Jr., Materials Science and Engineering. John Wiley, New York, 1985.

A. Eisenberg, The Glassy State and the Glass Transition. In Physical Properties of Polymers, J.E. Mark, ed. American Chemical Society, Washington, D.C., 1984.

J.D. Ferry, Viscoelastic Properties of Polymers, 2nd ed. John Wiley, New York, 1970.

M. Grayson, ed., Encyclopedia of Glass, Ceramics, Clay and Cement. John Wiley, New York, 1985.

H.H. Kausch, Polymer Fracture. Springer-Verlag, Berlin, 1987.

Standard for Polymeric Materials—Long-Term Property Evaluation, UL 746B. Underwriters' Laboratories, Northbrook, IL, 1986.

E.A. Turi, ed., Thermal Characterization of Polymeric Materials. Academic Press, New York, 1981.

9

Long-Term Properties of Polymeric Materials

VISCOELASTIC BEHAVIOR

Material properties data sheets list test results that have been gained during experiments extending over a time span of minutes or hours. For most structural materials these results are adequate since they also remain pertinent even if the part has to stand up under stress for a long time, months or years. But this is not the case for most plastics.

Since the melting temperature of all thermoplastics is rather low, these polymeric materials—though solid—retain some of their chain segment mobility down to the second-order transition temperatures, T_β, which could range below any ambient temperature on earth. In the preceding chapter reference was made to the high viscosity amorphous polymers attain at their glass transition temperature. At still lower temperatures their viscosity becomes so high that after the initial truly elastic deformation following the application of a load, the observation of any changes can be noticed only after a very long time.

All materials will react on the incursion of a tensile load by an instantaneous elastic elongation in the direction of stress. This reaction can be compared to a mechanical model represented by a spring. The stretching will be proportional to the applied force. If this applied force is maintained on plastic parts for a long period of time, the extension increases commensurately to the duration of loading because of the small extent of retained viscoelasticity. This viscosity can be compared with a mechanical model represented by a dashpot. Although the elastic component of this reaction is distinctly related to simple physical laws, the viscous component depends on many circumstances, which are further complicated by the multitude of chemical and structural variations in the polymers. Furthermore the dashpots also contain a certain friction component.

CREEP PROPERTIES

By combining a spring with a dashpot in either series or parallel connection, the experimentally obtained curves could not be reproduced satisfactorily. Over the years a large number of models with an increasing number of units

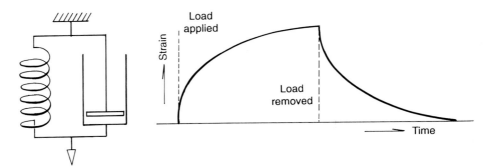

FIGURE 9.1. Voigt–Kelvin model and the strain–time viscoelastic creep response obtained according to this model.

have been described, but still no formula has been obtained that would reliably foretell the outcome of every experiment. Still the Voigt–Kelvin model shown in Figure 9.1 correctly describes the principle of viscoelastic substances, namely that any response will be based on the combined influence of an elastic and a viscous component. When one regards the multitude of positions and orientations of chain segments in an entangled agglomeration of polymer chains, it really becomes clear that the relationship between each neighboring segment pair will be somewhat different. One pair may be prevailingly spring dominated, whereas an adjacent one is more of a dashpot nature. These pairs may cover the whole spectrum from 99% spring and 1% dashpot to 1% spring and 99% dashpot. An abbreviated form of this description is shown in Figure 9.2. For simplicity, all four viscoelastic elements are shown at equal magni-

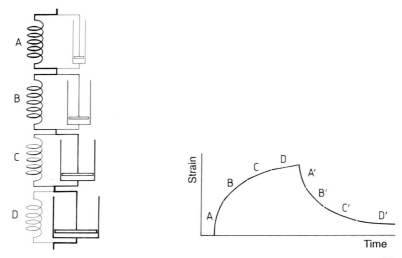

FIGURE 9.2. Abbreviated model for describing any viscoelastic response. The marked letters along the strain–time curve refer to the respective Voigt–Kelvin elements that dominate the shape of the curve although their influence is spread over a much broader range.

tudes, expressing 25% of the response. In reality each spring–dashpot pair could have a size of its own. This model could universally be applied by indexing its elements, as a sound is broken down into distinct intensities attributable to respective frequencies covering the whole audible spectrum. Since temperature affects mobility of the chain segments, increaes in ambient temperature will markedly lead to much higher creep distortions.

The determination of the creep behavior of materials can be established either on tensile specimens or on bars that are loaded in the center and their deflections followed. Additional information can be obtained by observing the recovery of the specimen after the load is removed. Figure 9.2 illustrates the shapes of creep responses that are observed on amorphous polymeric materials. A steady load is applied at time 0 whereupon an instant elastic strain becomes apparent. From there, the rate of creep can be inferred from the steepness of the slope. Depending on the extent of cross-linking, further creepage may come to an end after some time. Other specimens, especially those of low molecular weight or those tested at elevated temperatures, may continue to creep until they fail.

At the time the load is removed—as is illustrated in Figure 9.2—a large part of the elastic deformation may recover instantly (A'). Other parts, which have required a long time to deform, will be delayed also in their recovery over an extended time span (B' and C'). Since every dashpot must be considered to contain an appreciable amount of friction, recovery will always lag behind creep in regard to respective time intervals. The creep occurring under D may not recover at all but may remain manifested as a permanent set. To regain the original undistorted shape it will be necessary either to apply a force in the opposite direction or to heat the unloaded part above the glass transition temperature.

The ease with which such deformations will occur depends not only on the mobility of the chain segments (mainly their temperature) but also on the strength of the secondary bonds. A chain rupture under creep loadings is very unlikely unless other influences (aggressive chemicals such as ozone cracking of rubber) are also present. Therefore cross-linked three-dimensional network polymers with high cross-link densities are most resistant to creep as shown in Figure 9.3.

Polymers with strong hydrogen bonds (nylon) and polar bonds (polyvinyl chloride) are more resistant to creep than the polyolefins. Crystallinity will also help, but is not a total cure, since semicrystalline plastics still contain amorphous regions that are subject to creep. Probably most advantageous are the incorporations of rigid rings as chain members. In Chapter 8 it was shown how these types of polymers improve the thermal capabilities of plastics and the same is applicable to creep properties. Figure 9.4 indicates how two chains formed of single atom members can slip past each other in a caterpillar fashion (concertina locomotion), whereas a much higher energy is required to move six ring atoms only one step in unison past each other.

In many instances, time-extended creep curves will show four distinct sections as illustrated in Figure 9.5. After the instantaneously occurring deformation (combining an elastic and plastic share) subsides, a relatively large

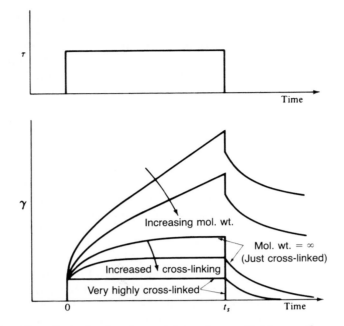

FIGURE 9.3.   The effect of molecular weight and cross-linking on the creep response of an amorphous polymer. Courtesy Stephen Rosen, Fundamental Principles of Polymeric Materials. John Wiley, New York, 1982.

amount of motion takes place in regions where many chain segments are still relaxed. In the curve's center section a more or less steady rate of creep becomes established as a result of the fact that most chains now consist of a uniformly strained material. Once weak areas emanate, the rate of creep again increases and the occurrence of failures accelerates.

To state creep properties as briefly as possible, the apparent modulus of elasticity expression is being used. If a test specimen is subjected to a specific

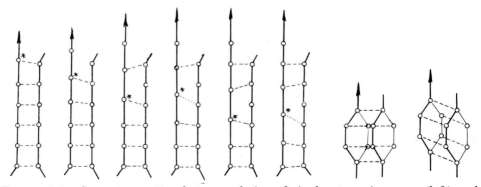

FIGURE 9.4.   Creep locomotion between chains of single atoms in a row (left) and resistance to sliding past each other with rigid, six atoms containing rings (right). From G. Gruenwald, Kunststoff Rundschau 11, 525, 1964.

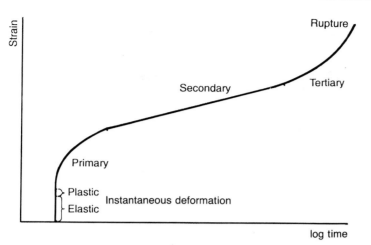

FIGURE 9.5.   Extended creep curve of a typical thermoplastic.

load at a certain test temperature, the rapidly occurring extension will be indicative of its modulus of elasticity. This becomes expressed by the initial slope of the stress–strain curve. If that initially applied stress is maintained at the same level for a long time, a noticeable further extension will occur in case of creepage. Connecting the latter point with the origin of the curve will effect a shallower slope representing the *creep (apparent) modulus*. Creep test data are becoming available with increasing frequency. The *Modern Plastics Encyclopedia* regularly compiles them in a sequence of tables. Still, these data can be used only for comparison reasons since deviations in temperature, applied stress, and duration of test must also be considered.

The knowledge about creep properties of plastics is of great importance where plastic parts are expected to remain structurally stable in spite of long duration loadings. In this regard the realm of plastic pipes has attracted special interest. A problem arose in ensuring the serviceability of those pipes for 50 years without being able to extend testing for such a long time span. Figure 9.6 shows how creep tests to failure were speeded up by raising the test temperature. When a large number of pipe sections were subjected to internal pressure at different stress and temperature levels the resulting failures (leaks) were charted on graphs. On double logarithmic paper the failures appeared first on straight slopes, making it seemingly possible to predict the time of failure for the samples still pressurized. All these initial failures could be classified as ductile breaks. Soon after a second mode of failure emerged (brittle failure), which occurred at much higher rates (steeper slopes). Though both failure modes are characterized by different slopes, the presently available information makes it possible to obtain with high probability an affirmation of the suitability of any material.

Very short time experiments at elevated temperatures can be employed to obtain only quick quality assurance about the extrusion process. For the evaluation of changes in compounding or in the polymer itself, a much more

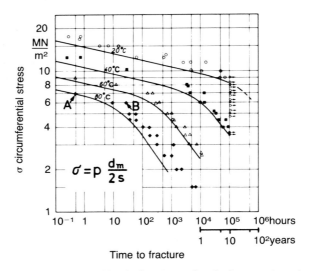

FIGURE 9.6.   Time to failure of high density polyethylene water pipes. Courtesy H.H. Kausch, Polymer Fracture, Springer-Verlag, Berlin, 1987.

extended test duration is required since extrapolation of data must always be appraised with caution. Extending observations over 1% of the expected life time is a must, although 10% is recommended.

STRESS RELAXATION

Whereas Figure 9.3 illustrated the creep response for amorphous polymers, Figure 9.7 illustrates the general stress-relaxation behavior of those plastics.

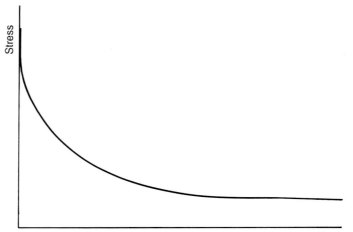

FIGURE 9.7.   Stress relaxation with time curve.

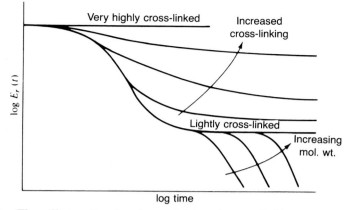

FIGURE 9.8.   The effects of molecular weight and cross-linking on stress-relaxation master curves. Courtesy Stephen Rosen, Fundamental Principles of Polymeric Materials. John Wiley, New York, 1982.

In this instance too, data can be represented in log/log graphs as apparent modulus of elasticity values (Fig. 9.8). Stress relaxation is mainly of importance where tightness of joints of plastics assemblies is of utmost importance or where plastics are used for gaskets and seals. Requirements are becoming especially severe when applications also involve wide temperature variations. Because of the high coefficient of thermal expansion of plastics, strain may become intensified at the high temperature side leading to increased stress relaxation.

Occasionally good use is made of stress relaxation in plastics assemblies when it can be ensured that residual stresses will not pose any problems. Under force combined parts can be made to conform to a desirable shape. An additional annealing process will ensure that the impressed alteration in shape becomes permanent. The elastic strain response under stress relaxation conditions is also different from the response under creep. Since here most of the imposed stress is curtailed by creep, the instantaneous elastic strain recovery will be only minimal. Another portion is regained—in time—by reverse creep.

TIME–TEMPERATURE SUPERPOSITION

The similarity of the effect both time and temperature may have on mechanical properties within the viscoelastic range of plastics has led to the adoption of the time–temperature superposition principle. In Figure 9.9 the whole viscoelastic spectrum of an elastomeric polymer, polyisobutylene, is shown. In this case it would have been impossible to obtain all the stress–strain data at room temperature because the experimental time would have to include durations of less than 1/1000 sec. However, by conducting the tests at a number of spaced temperatures [$-113°F$ to $122°F$ ($-80.8°C$ to $50°C$)], sufficient data

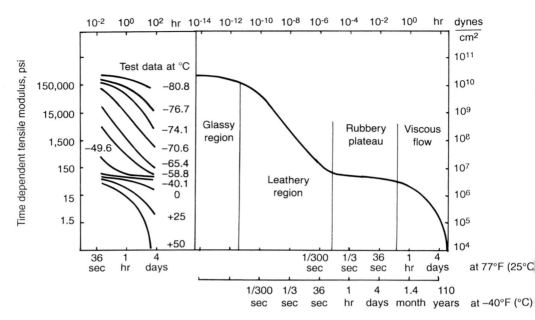

FIGURE 9.9. Time–temperature superposition for polyisobutylene. At the left, the experimentally determined apparent modulus values that were obtained individually at the temperature and test times indicated are entered. At right, the master curve for 77°F (25°C) was composed by superposing the data under consideration of the respective shift factor, a_T, for each test temperature. Adapted from data by A.V. Tobolsky and E. Catsiff, J. Polym. Sci. **19**, 111, 1956.

points could be obtained within a reasonable time span from 0.5 min to 4 days. Utilizing an experimentally determined shift factor, plotted in Figure 9.10, the individual curves can then be combined into a single composite master curve. The undulated shape of the latter curve attests that the general behavior of all amorphous thermoplastics, whether elastomers or high-temperature rigid plastics, will encompass the same four behavioral regions only at different temperature zones. Once a relationship for a certain material has been established, required design data can be reliably extrapolated utilizing more conveniently obtainable test conditions.

The extent of the rubbery plateau is very much dependent on the molecular weight of the polymer. The steep changes in modulus occur only in those regions where intermolecular bonds rapidly vanish, that is, at the glass transition temperature and the melt temperature. In case of cross-linked linear polymers the latter decline is eliminated because of the presence of strong covalent cross-linking bonds. The typical network thermoset plastics, which are tightly cross-linked, will not even show the first decline in modulus.

DYNAMIC MECHANICAL PROPERTIES

Because of the viscous component in the stress–strain behavior of most plastics materials, an additional effect can be observed when plastic parts are

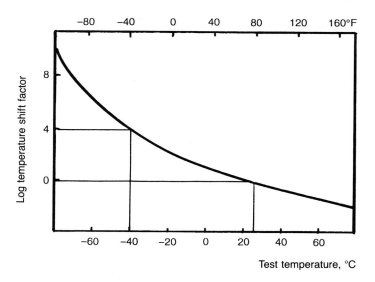

FIGURE 9.10.   Shift factor utilized to establish master curve.

subject to cyclic loading. Truly elastic materials, such as most metals, do not require special considerations. In plastics there will be a time delay between the cause (application of a cyclic load) and the effect (deflection of the part), similar to the observations on alternating current systems between voltage and amperage. The mobility or lack of mobility at the atomic level of the polymer links must again be taken into account. In addition to the chemical composition, both the frequency and the temperature will greatly influence the results. One must envision that under dynamic loading conditions the applied force symbolized by the complex modulus of elasticity is partitioned into two 90° diverted entities. The real component, generally called storage modulus, represents the part of the energy that has been just temporarily stored and will be regained at the end of each cycle. The other component, called loss modulus, represents that fraction of the energy that will be mechanically lost but converted into internal frictional heat. The value for the vector that proportions the two energy components can also be expressed by the tangent delta or loss angle.

Although the effect of loading on deflection is generally less under dynamic loading than under a steady load of equal magnitude, the system will mainly be indirectly affected by the energy losses, which are converted to heat in the material. The rising temperature of the part may exceed the tolerable temperature limit for the plastic used. The ensuing loosening of the intermolecular structure of the polymer becomes especially noticeable with lower molecular weight resins. The observed mode of failure becomes increasingly expressed in early occurring brittle fractures. On the other side, the damping effect based on the energy absorption can become beneficial where vibration damping (noise reduction) is of interest.

Hardness seems to dominate the effect on dynamic properties of metals. Spring steel exhibits a near zero loss modulus whereas the loss modulus of lead is very high. This rule is not followed with polymers, which are influenced by temperature and frequency changes to a greater extent. The outstanding materials in regard to low loss modulus values are soft, vulcanized natural rubber, silicone rubber, and stiff fiber-reinforced composites.

Unquestionably any of the dynamic testing procedures, not only dynamic mechanical but also electrical and optical testing, have contributed significantly to the understanding of many minute structural features in plastics not obtainable any other way.

Fatigue Strength

The fatigue strength of a plastic is determined by stressing specimens at various stress levels, frequencies, and amplitudes until failures occur. In most cases the endurance limit will be reached if a test specimen survives a few million cycles. Figure 9.11 shows SN curves (stress versus number of cycles) for different polystyrene polymers. This also illustrates how important it is to select high-molecular-weight plastics when exceptional first-rate properties are demanded.

The Metastable State of Thermoplastics

As has been discussed in previous chapters, in amorphous thermoplastics the transformation from a liquid to a solid is not an instantaneously occurring event since certain sections of the polymer chains become tacked to each other

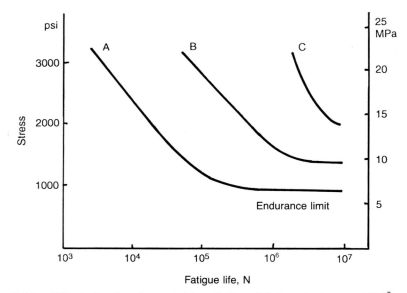

FIGURE 9.11.   Effect of molecular weight on fatigue life in polystyrene. (A) $\bar{M}_w = 1.6 \times 10^5$, (B) $\bar{M}_w = 8.6 \times 10^5$, and (C) $\bar{M}_w = 2.0 \times 10^6$. From J.A. Sauer and G.C. Richardson, Int. J. Fract. **16,** 499, 1980.

before the thermodynamically aspired overall order can be established. At the glass transition temperature, the viscosity is so high that interchain mobility ceases on further cooling. The relatively high volume contraction expected to occur with the remaining mobile sections becomes restricted. Instead a much lower linear coefficient of thermal contraction, characteristic for the ordered, rigid component, is observed. Therefore, a certain amount of free volume will become locked into the system, preventing a better utilization of space by the chain segments. Equilibrium at temperatures 20°F (10 K) or more below the glass transition temperature cannot be obtained but only approached by annealing for an extended period of time. Ensuing cooling to room temperature, however, again restrains attainment of equilibrium. Properties of rather fast cooled parts, as it becomes a necessity for production items, will contrast with those of well-annealed parts. The latter might show an increased specific gravity, a lower mobility of some of the chain segments leading to a higher modulus of elasticity, and an accompanying lower potential extensibility, which results in lower impact properties. The annealed form will also have a greater hardness, higher rigidity, and better electrical resistivity. In many instances the as-molded condition appears to represent the norm since at ambient temperatures no changes become apparent even after an extended period of time. In other cases the metastable state is transitory and can be obtained only under controlled circumstances. For example, several patents have detailed how to chill polyvinylidene chloride rapidly so that the soft sheet and film material can easily be cold formed. Thereafter at room temperature they quickly stabilize by conversion to a hard and rigid semicrystalline product.

This phenomenon is now generally described as physical aging. For a material to exhibit physical aging, it must contain segments in its structure that have the potential to realign in a more ordered arrangement. Highly crosslinked, rigid materials and rubbers are not prone to such a change.

The conversion from a metastable state to a more stable state is not equally pronounced in various amorphous materials. The greater the temperature difference between the annealing temperature and glass transition temperature, the more time is required to approach the state of ultimate stability. An overly narrow temperature gap will quickly lead to an equilibrium; however, it will relate only to that relatively elevated temperature. It still cannot procure the stable form for a lower ambient temperature condition. Depending on the circumstances and the intended application, the more stable form may be considered advantageous or disadvantageous. An early publication by a polycarbonate producer recommends annealing for molded parts to improve their stiffness and modulus of elasticity. Now, the concurrent loss in impact properties of such high impact-resistant thermoplastics (polycarbonate and polysulfones) is becoming more of a concern. For most thermoplastic materials these changes in properties can be clearly proven only by very sensitive dynamic mechanical tests. Large differences are found in impact strength test results when very high-energy-absorbing ductile specimens convert to specimens exhibiting brittle failure, producing much lower impact strength values. However, it should be emphasized that these brittle fracture levels are still considerably higher than those found in typical brittle materials.

It is interesting to note that this aging can be nullified by briefly heating the part back to the glass transition temperature. On the other hand, an extended loading, which would result in a slight creep deformation, may evoke an aging effect similar to annealing.

It is of importance to realize that physical aging should not be mistaken for the generally used aging process that causes a chemical deterioration of the polymer molecule. In the case of polycarbonate it has been observed that Izod impact bars if heated for several days at 255°F (125°C) result in relatively low brittle failures on impact testing whereas if heated for weeks at 300°F (150°C) these bars will maintain their strong, high ductile failure mode.

This sluggishness of the polymer chains to attain a more stable structure at ambient temperatures can also be overcome by increasing the mobility of polymer segments by the addition of plasticizers or solvents. When it was recognized that small amounts of plasticizers in polyvinyl chloride will produce a result opposite of what was expected, the designation of antiplasticizer was introduced. The highest tensile strength values of 9000 psi (62 MPa) were obtained with polyvinyl chloride specimens containing 10% tricresyl phosphate plasticizer whereas the strength values both for the unplasticized specimens and those containing 18% plasticizer reached only 6000 psi (41 MPa). This anomaly extends to many other properties: density, modulus of elasticity, elongation, impact strength, refractive index, and electrical conductivity. In all cases, extrapolation of property values to 0% plasticizer content invariably would lead to a much denser, stiffer, and stronger neat polymer than is actually encountered. The explanation for this is that the small amount of plasticizer—similar to annealing—sufficiently mobilizes the polymer chains so they can arrange themselves in a more ordered structure that requires relatively less space.

This phenomenon also leads to other complications that have practical consequences. The presence of low-molecular-weight additives used as dispersants or compatibilizers in colorants or other concentrates may surprisingly lead to brittle fracture of molded specimens. However, chemically noncompatible dispersants, such as many lubricants and waxes, might not affect those plastics properties.

Similarly, solvents have been documented to cause the loss of ductile impact behavior in certain plastics. This has been seen in cases where solvent-based adhesives are employed for assemblies or where coatings containing aggressive solvents are used. The mere removal of the absorbed solvent is not sufficient to reverse this embrittlement. Ductile failure modes can be regained only when the physically aged polymer structure is reconverted to the metastable form by an additional brief heating of the plastic part to the glass transition temperature.

REFERENCES

G. Gruenwald, Lightly Plasticized Plastics. Kunststoffe **50,** 381–387, July 1960.

Standard for Polymeric Materials—Long-Term Property Evaluation, UL 746B. Underwriters' Laboratories, Northbrook, IL, 1986.

L.C.E. Struik, Physical Aging in Amorphous Polymers and Other Materials. Elsevier Scientific Publishing Comp., Amsterdam, 1978.

L.C.E. Struik, Physical Aging. In Polymers: An Encyclopedic Sourcebook of Engineering Properties, J.I. Kroschwitz, ed., p. 36. John Wiley, New York, 1987.

10

Chemical Properties of Polymeric Materials

SOLUBILITY

A slow dissolution or disintegration represents a threat to many structural materials. In the case of metallic and mineral materials, water as a carrier for acidic or alkaline reagents can destroy some of them in time either by corrosion or erosion. These processes are based on chemical reactions. Structural polymeric materials, both the synthetic polymers and the structural biopolymers (cellulose), are not dissolved by water or any slightly acidic or alkaline solutions. Only a few polymers, because of the presence of large numbers of hydrophilic (e.g., hydroxyl, amino, and carboxyl) groups, become water soluble. Since water is one of the most polar solvents, it is understandable that it can dissolve these highly polar groups containing polymers.

Polyethylene and the other polyolefins—at the other extreme—contain only nonpolar alkane groups, making them totally insoluble in water and also completely nonabsorptive for water. The rule that solvents composed of certain chemical groups will dissolve those substances containing the same or chemically similar groups applies to the polyolefins only to a certain degree. The reason for this is that the dispersive forces in such nonpolar solvents (aliphatic hydrocarbons) are so weak that solutions can be obtained only at high temperatures.

Many solvents lie between these two extremes and represent characteristic classes of their own with specific solvent capabilities. Some of them are the aromatic solvents (containing the benzene ring), the halogenated solvents (which may vary greatly in their polarity, depending on the kind of symmetrical positioning of the halogen atoms), the ether-type solvents, and the ester and ketone solvents. Depending on the character of these functional groups, each of these solvent molecules will have a certain cohesive energy density, which means a very specific strength relating to the intermolecular forces acting between these molecules. Chances for dissolution are best when both the solvent and the solute (polymer) have the same cohesive energy density, enabling the solvent molecules to wedge themselves between touching polymer chains. When the cohesive energy density cannot be matched by a single solvent, a mixture of two solvents might satisfactorily dissolve the polymer. This may especially be necessary for dissolving copolymers and blends. For the production of cast films these solvents are sometimes replaced by mixtures of chlorinated solvents to avoid an explosion hazard.

Some examples for polymer solubilities should be mentioned:

1. Water-soluble polymers include polyvinyl alcohol, polyethylene oxide (but not the highly crystalline polyoxymethylene or acetal polymers), cellulose methyl ether, hydroxyethyl cellulose, carboxymethyl cellulose, polyvinyl pyrrolidone, polyacrylic acid, and acrylic acid copolymers.
2. Toluene as an aromatic hydrocarbon represents the best solvent for polystyrene.
3. Ester and ketone solvents are preferred for cellulose esters, acrylic esters, and many other lacquer resins.

Low-molecular-weight substances which readily dissolve in solvents, are soluble up to high concentrations, and affect viscosity of the solutions only at quite high concentrations. Solubility of high-molecular-weight polymers is equally dictated by the polymers' cohesive energy densities, which have been established and published in handbooks. However, even under favorable conditions the necessity to cleave the large number of secondary bonds connecting neighboring long chains retards the dissolution process. Low-molecular-weight molecules are quickly carried off by the low-viscosity solvent. With polymeric materials, the same solvent will first swell the surface layer to separate one stretch of the polymer chain from adjacent polymer chains and then work its way along the whole length before the polymer can be dispersed in the solvent. Already low concentrations of polymer will increase viscosity of the solution causing further delay of the solvating process. These observations must be heeded when polymeric resin powders are used to prepare solutions. These powders must first be completely wetted with nonsolvents or cosolvents before the good solvents are added. Otherwise lumps requiring hours to dissolve will be formed.

Since no covalent chemical bonds are severed by solvents, thermoset polymers, if they form tightly intertwined networks, such as the phenol-formaldehyde resins, will not be affected by them. Neither are the polymers with more spread out cross-links, as present in rubbers, elastomers, and polyurethanes, soluble in solvents. But since the cross-links connect long entwined chains, the latter may stretch out, allowing the absorption of large amounts of solvents. The solvent that best matches the cohesive energy density of the polymer results in maximum swelling. Some rubbers can swell several hundred percent in volume and, as a result, lose all their strength and abrasion resistance. Since natural rubber and some synthetic rubbers belong to the family of aliphatic hydrocarbons, they are sensitive to gasoline and lubricating oils. Only polymers and elastomers that contain chlorine, fluorine, sulfur, or nitrile groups are resistant to swelling by these aliphatic hydrocarbon solvents and oils.

In semicrystalline polymers, certain sections of the polymer chains are deposited within the rigid lattice of crystallites. Therefore most of them are unaffected and insoluble in common solvents unless the secondary bonds acting between polymer chains are weakened by increasing the temperature.

Although the designer might initially be more concerned about the effects solvents have on plastics, solvents and solutions of plastics have found many practical applications. Volatile solvents are used to bond many amorphous plastics to themselves. Nonvolatile solvents are incorporated as plasticizers in certain plastics to make them flexible. Solutions of many polymeric resins represent coating materials such as paints or varnishes. In this way the excellent environmental resistance of polymeric compounds can be utilized to protect the surfaces of ferrous metals or seal the porous structure of wood. The most important resins for those applications are the phenolics, alkyds, cellulosics, acrylics, silicones, and polyurethanes.

WATER AND MOISTURE SENSITIVITY

Similar to the action of oil toward rubber, water molecules will penetrate into polymers even if they are not water soluble. Since water is inevitably present in the atmosphere, although at quite varying concentrations, this effect must always be taken into consideration. On the other hand, metals and ceramics can accommodate water molecules only at their surfaces. Only the polyolefins, consisting exclusively of hydrocarbon chains, do not absorb any water. Still, very loosely arranged hydrocarbon polymers, such as polymethylpentene and rubber, may have quite high water vapor transmission rates. The polyamides (nylon resins) can reach the highest water absorption values of up to 10% under saturation conditions. Such a high concentration will naturally have a noticeable effect on that material's properties and on the dimensions of molded nylon parts. In this case the water can be considered a plasticizer that causes significant reduction in the modulus of elasticity, tensile strength, and hardness values. Conversely, the water will increase elongation and improve impact properties. Since water absorption represents a rather slow process, those extremely high saturation values are practically never obtained unless those parts are continuously under water. Even under widely fluctuating relative humidity conditions, the water content of most polyamides will seldom exceed 1%, which is just about the range in which optimum mechanical properties are obtained. Frequently two sets of property values for nylon resins are listed relating to different moisture contents. Exceptionally dry nylon resins would be too brittle for many applications. To avoid this condition, bone dry, just molded parts are occasionally exposed to boiling water or salt solutions for a certain time to obtain the appropriate water content from the start.

The water absorption of all other engineering plastics is not high enough (much less than 1%) to cause any problems in regard to dimensional tolerances or mechanical properties. Only the acetal polymers and the cellulosics reach the 1% level at saturation. Plastics with lower than 0.1% water absorption include, besides the polyolefins, polystyrene, the fluoropolymers (polytetrafluoroethylene), and polyvinylidene chloride. Only occasionally will extended submersion in water lead to turbidity or to slight weight losses, and then primarily only when water-soluble salts or detergents are left in the plastic compound.

Still, these small amounts of absorbed water may occasionally have an impact on processing, as illustrated in the following three examples:

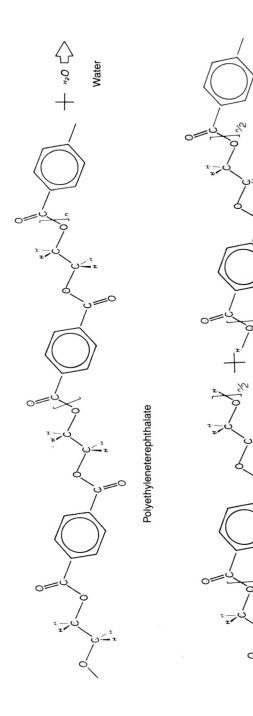

Polyethyleneterephthalate

Low-Molecular-Weight Polyethyleneterephthalate

Water

(1)

1. During injection molding of not quite dry compounds, water vapor may condense on the cooler mold surface causing the molded part to display surface streaks.
2. When plastic sheets are quickly heated to the plastic's softening range, the steam forming internally from the absorbed water may have no time to slowly diffuse through the surface but will form numerous bubbles, making the sheet unsuitable for thermoforming. Time-consuming predrying will become necessary.
3. The most serious effect of moisture can occur when hydrolysis-sensitive plastics (all polyesters, polyurethanes, polycarbonate, polyamides, and polyimides) are heated in the injection cylinder in the presence of traces of water. Because of the vast difference in molecular weights between water (18) and the polymer (in excess of 50,000), a concentration of only 0.035% water can potentially cut the molecular weight of a polyester resin in half, see chemical formula (1). This material degradation cannot be reversed.

Since the uptake of water represents a slow process, the drying of plastic pellets or sheets also becomes a time-consuming process. Drying temperatures cannot be raised to desirable levels because of the softening and coalescing of the plastic pellets. Therefore, air of extremely low moisture content, a dew point below 0°F (-18°C), must be employed to obtain satisfactory results when drying moisture-sensitive plastics.

AGGRESSIVE CHEMICALS

Although many plastics are derived from coating materials in which they were functional in protecting ferrous metal and wood structures, they themselves are not totally spared from the ravages of environmental influences over time. As coatings, applied only in thicknesses of less than 1/100 in. (1/4 mm), they will usually last for several years. Since plastics parts are always much heavier in cross section, wear or erosion is seldom a problem for them. However, there are certain aggressive chemicals that must be kept away from their surface.

All important plastic materials are established on carbon atom chains with pending side groups. They are therefore subject to oxidation by strong oxidizing agents, causing destruction of most types of plastics. Strong alkaline reagents must be listed in second place, followed by strong acids. Many times volatile reagents of those classes of materials will attack plastics much faster because of their more rapid penetration (e.g., ammonia versus sodium hydroxide). Plastics exhibiting resistance toward certain chemicals have been mentioned in other sections. The influence of chemicals over extended time spans or in combination with other stresses will be dealt with in the following section.

What should be emphasized here is that most plastics represent a compounded product consisting of diverse materials. Therefore losses in its properties may be caused just by the destruction or elution of additives such as plasticizers, stabilizers, or fillers, rather than attacks on the polymer itself.

ENVIRONMENTAL STRESS-CRACKING

The phenomenon commonly called environmental stress-cracking represents one example in which a chemical agent by itself does not affect the properties of a plastic, but may cause brittle fracture to occur early when a plastic is simultaneously subjected to mechanical stresses containing a tensile force component. In a rigorous way, these incidents may represent the rule rather than the exception. Improvements in tensile properties of many materials, including metals, ceramics, and glasses, have repeatedly been reported when testing is done in vacuum excluding the presence of any gases and vapors present under ambient atmospheric conditions.

Here we will concentrate on notorious impairments in properties such as those first noticed by metallurgists on copper-containing alloys, which can fracture under low stresses in the presence of ammonia. Plastics will vary in regard to their susceptibility to environmental stress-cracking. Both the level of stress, perhaps better expressed as strain, and the nature and concentration of the chemical affect this process. Below a certain threshold strain (generally up to 0.2%), the appearance of cracks or crazes will be delayed infinitely. Also certain correlations exist between the chemistry of the plastic and the chemistry of the reagent. Some of these perilous pairs are polyethylene and organic acids or detergent solutions, nylon and acids or certain salt solutions, polycarbonate and ketone or chlorinated solvents, and polyphenylene oxide and hydrocarbon solvents.

Although no clear relationships can be established, completely inert substances seldom cause problems. Chemicals that readily dissolve plastics may not cause cracking since these solvent molecules can penetrate the surface and swell the plastic, resulting in reduced surface strain and retarding further penetration of the solvent. If the solvent environment does not persist over an extended time because of evaporation, as found in the case of paint or solvent bonding of molded parts, no detrimental effect will be noticed. As an example polycarbonate and methylene chloride, a good solvent, should be mentioned. If the good solvent is replaced by an inferior solvent such as carbon tetrachloride, the poor solvent is capable of severing van der Waals bonds between adjacent polymer chains only where chains are under strain. Since this solvent is not absorbed by the plastic, the incipient crack remains open and the fissure may propagate extremely rapidly through the whole part. Environmental stress-cracking may occur rapidly within a fraction of a second but could also be delayed for hours and years.

That environmental stress-cracking involves only the separation of secondary bonds (not the covalent bonds of the chain molecules) can be concluded by the complete resistance demonstrated by the ultrahigh-molecular-weight polyethylene plastics. The same is also true of cross-linked polyethylene. In addition to increasing molecular weight, the incorporation of elastomeric segments to reduce the modulus of elasticity can lessen or at least delay the occurrence of environmental stress cracks. Similarly, avoiding the formation of frozen-in stresses or reducing them by annealing presents a beneficial effect.

PERMEABILITY AND BARRIER PROPERTIES

The food industry requires packaging materials that are impervious to gases, vapors, and aromas. Since glass and metals are totally impervious to any other substance, they have provided ideal packaging materials for centuries. However, their high cost and weight have limited their use. When the lower cost materials, such as paper, cellophane, and later plastics, became available, they were used as replacements. Plastics for packaging exhibit further advantages in regard to simplified processing and lower energy requirements. But it became evident that all organic substances, including plastic film and sheet materials, permit gases (oxygen, carbon dioxide, water vapor, etc.) and liquids (water, oils, flavoring agents, etc.) to permeate them to a certain degree.

The aim is in most cases to reduce

1. the permeation in both directions to negligible levels,
2. the loss of any food ingredients, and
3. the migration of undesirable substances from the environment, while preserving the wholesomeness of the packaged goods.

Only in some instances is the permeation of substances desirable (e.g., to retain the fresh red color of meat, oxygen must be able to penetrate the packaging film).

To select the appropriate packaging material it is necessary to understand the whole permeability process, which can be subdivided into four separate steps. First, the migrating material must be absorbed by the plastic material at the surface. Next, the solubility constant will determine how much of the substance can be picked up. Third, the substance must diffuse through the packaging material. And finally, it must be desorbed at the opposite surface. In many cases these four reaction rates could be quite different. The overall performance is dictated mostly by the magnitude of the slowest reaction rate.

Polymers containing many highly polar groups, such as the hydroxyl groups in polyvinyl alcohol or uncoated cellophane, represent excellent barriers for gases, but they are not only very poor as barriers to moisture and water vapor but also suffer great losses in their gas barrier properties when wet or swollen by water. The much higher values obtained under these circumstances are given in Table 10.1 in parentheses. High-density polyolefins, on the other hand, are excellent moisture and water vapor barriers but poor gas barriers. Barrier properties that can be regarded as fair overall are obtained from plastics when they contain some polar groups such as nitrile, chlorine, fluorine, or ester groups. Further beneficial factors include high chain stiffness, a high glass transition temperature, and a close chain-to-chain packing evidenced by a higher than expected specific gravity.

Relating to these statements, the following examples should be cited. Increasing the crystalline content of plastics improves gas barrier properties. According to recent publications, highly crystalline and dense liquid crystal polymer films can surpass the barrier properties of the best established plastics products. The addition of pigments will also be beneficial, unless gaps

TABLE 10.1. Barrier Properties[a] of Plastics Materials

Polymer	Gas Transmission Rate at 1 atm and at R.T. ($cm^3 \cdot mil/100\ in.^2 \cdot 24\ hr \cdot atm$)		Moisture Vapor Transmission Rate at 100°F (38°C) ($g \cdot mil/100\ in.^2 \cdot 24\ hr$)
	Oxygen	Carbon Dioxide	
Polyacrylonitrile	0.01	1	1
Commercially available copolymers	0.7	1.6	4
Polyvinyl alcohol	0.01 (25)[a]	0.03	2000
Commercially available copolymers	0.05	0.1	4
Polyvinylidene chloride	0.1	1	0.1
Commercially available copolymers	1	5	0.5
Cellophane, uncoated	0.1 (200)	0.5	100
Nylon	2 (5)	10	20
Polyethyleneterephthalate, oriented	5 (5)	20	2
Polyvinyl chloride[b]	15	30	5
Cellulose acetate	100	1000	100
Polycarbonate[c]	250	1000	10
Polystyrene	300	1200	8
Polypropylene and high-density polyethylene	150	500	0.5
Poly-4-methylpentene[d]	4,000	10,000	10
Silicone	15,000	45,000	50
Polychlorotrifluoroethylene	10	20	0.03

[a] Values in parentheses have been obtained at 100% relative humidity conditions.
[b] Similar values for polymethyl methacrylate and polyoxymethylene.
[c] Similar values also for modified polyphenylene oxide.
[d] Other polyolefin types are situated in between: neoprene, butyl rubber, polybutene, low-density polyethylene, ionomers, and polybutadiene.

within the polymer and pigment interface will appear because of bond loss (improper surface treatment). Orientation will only be marginally beneficial. Its predominance in packaging materials is founded more on the need for imparting adequate toughness to the package. Anything that will soften the plastic becomes devastating, be it plasticizers, impact modifiers, or just an elevation in temperature. An increase of only 10°F (5 K) can increase permeability by roughly 50%. Also, the open structure olefins, especially polymethylpentene which has a specific gravity of only 0.835, will show much higher gas and especially moisture permeation than high-density polyethylene (0.95). This is also true to a certain degree for polypropylene (0.90). The open structure of polystyrene (specific gravity 1.04, the lowest of all benzene-ring-containing polymers) makes it also very permeable to gases and moisture. Polychlorotrifluoroethylene, with a specific gravity of 2.2, represents the best material in regard to water vapor permeability, but it is not the best for every gas. The size, shape, and polarity of the permeating substance has an effect also, as does the thickness (permeability is not necessarily proportional to thickness), void content, and environment around the film material.

The three best barrier film materials, polyacrylonitrile, polyvinylidene chloride, and polyvinyl alcohol, can be obtained only by costly solution casting processes. Therefore these materials are primarily used in the form of copolymers that can easily be melt extruded into films. Since packaged foods are in many instances stored only for a short period of time, very thin films of these materials provide sufficient barrier properties, making the mechanical durability of the package the limiting factor. For that reason three-layer coextruded films were introduced. At the outside relatively thick polyolefin, nylon, or polyethyleneterephthalate layers provide the necessary strength. One type of ketchup bottle now in use contains only two very thin [0.5 mil (0.01 mm)] ethylene–vinyl alcohol copolymer layers as barriers. More than 98% of the weight of this bottle consists of polyethyleneterephthalate. Since in this case no adhesive layers are employed, a noncontaminated polyester can readily be segregated and reprocessed for recycling.

Increases in the number of layers to five and seven have been promoted to furnish one or more of the following benefits:

1. improving adhesion between dissimilar materials by using tie-layers;
2. protecting the gas barrier layer (polyvinyl alcohol copolymer) from absorbing moisture; and
3. providing a heat-sealable inner surface.

As in plywood, an odd number of layers is chosen to prevent warping or curling of films.

The barrier properties of some important plastics materials are listed in Table 10.1. In the literature quite diverse numbers are sometimes found. Still, in regard to material selection, they all represent useful data even if only the placement of their decimal points is considered correct. This becomes clear if one, for example, has to establish whether a sufficient amount of carbon dioxide must be retained in a container for 3 days, 1 month, 1 year, or 10 years, where each step represents one order of magnitude.

In most cases the permeability for nitrogen is only about one-quarter of that for oxygen. Not quite so consistently the permeability numbers for carbon dioxide are in the average about six times those of oxygen. An old competitive packaging film, consisting of surface-coated regenerated cellulose, called cellophane, should be mentioned here. Its excellent barrier properties, especially when coated with polyvinylidene chloride, would give it first preference if it were stronger and if it could be formed.

All rubber compounds are high in gas permeability but silicone rubber tops them all. The exceptionally low moisture vapor permeability of poly-chlorotrifluoroethylene has earned it many applications in the chemical industry and the field of medicine.

Several processes using only surface treatments to improve the barrier properties of plastics should also be mentioned. The permeability of gasoline through polyethylene fuel tanks can be drastically reduced by treating the inner surface. This can be accomplished either by sulfonation, posttreating with sulfur trioxide followed by rinsing with ammonia, or by fluorination, adding fluorine gas during the blowing process. Because of their shortcomings in regard to permanence, these processes are falling into disfavor, giving preference now to coextrusion forming processes. The chemically converted surface will hinder the absorption of gasoline. Metallizing polyethyleneterephthalate or oriented polypropylene films represents another way to reduce gas transmission rates, which also provides a glossy, light reflecting surface. By a plasma-enhanced silica vapor deposition process, polyolefin and polyester films and bottles can receive a quartz-like film coating that primarily reduces moisture permeability while maintaining clarity. The extremely small amount of silica deposited has no effect on recycling these products together with pure resins.

By the addition of 5% specially modified nylon resin, a polyolefin compound can be obtained, the microstructure of which exhibits thin lamellae of the incompatible nylon throughout the bulk of the plastic. The escaping gasoline molecules must traverse a tortuous path, retarding permeation considerably.

Figure 10.1 illustrates the application of barrier materials in one specific area; this should serve as a reminder that—not only in the case of packaging—it is increasingly difficult to develop materials so improved that they render the previously used materials obsolete.

Polymer Degradation

All processes that lead to drastic deterioration of any of the properties deemed essential for a specific application can be included under the term polymer degradation. These changes will be manifested in most cases by a bond scission reaction in the macromolecule, but they could also be caused by excessive cross-linking. These reactions can be initiated by a great number of influences including thermal, mechanical, photochemical, high-energy radiation, or biochemical effects. Once started, deterioration accelerates by more extended chemical interactive reactions.

FIGURE 10.1.   Presently coexisting beverage containers made out of metal, glass, cardboard, blow molded plastic, laminated plastic barrier materials, film fabricated stand-up container, and plain polyethylene pouch.

A mechanical scission of polymer molecules is of no concern during ordinary processing of plastics, although it may occasionally affect fibrous reinforcements. Still, this type of molecular breakdown must be mentioned since such a process is being utilized when rubber stock is subject to high shear in Banbury mixers or on rubber mills.

In all cases, even if the intent is to synthesize a photo- or biodegradable plastic, breakdown of the plastic during manufacture, storage, and use must be prevented. This is accomplished by purifying the polymer and all compounding ingredients and by selecting appropriate stabilizing additives. Further protection must be attempted by preventing or minimizing the above-listed causes for degradation. It must also be kept in mind that even the best stabilizers will—in time—be degraded or leached out of the plastic.

Photodegradation is a characteristic of nearly all organic materials, since the energy level of their chemical bonds is about equal to the energy of the near ultraviolet radiation. To accomplish an energy transfer, light at that wavelength must be absorbed by the plastic (either the polymer itself or a UV-absorbing additive). The acrylic polymers are known for their weather stability; they will not absorb the sun's UV radiation. Polyolefins initially will not absorb this radiation either; however, after undergoing a minor chemical oxidation, the newly formed radiation-absorbing carbonyl groups further promote the breakdown of the polymer. This reaction path has been utilized for the production of controlled photodegradable polyolefins by incorporating a small amount of carbonyl groups via copolymerization of ethylene with carbon monoxide. These plastic materials remain stable as long as they are kept indoors since the window glass will absorb all the active UV radiation of the sun. Vinyl ketones represent other monomers that can incorporate an active carbonyl group. They are mainly used for disposable polystyrene items. It should be mentioned that some substances, such as transition metal salts containing iron or nickel, which are employed as light stabilizers at higher concentrations, will—when used in low concentrations—also accelerate the polymer photodegradation, once it has started.

High-energy radiation causes degradation of the polymers by the interaction of strong electromagnetic radiation or particle radiation. Only some high-temperature resistant polymers remain less affected. Under any very-high-intensity radiation condition of short duration the reaction mechanism will shift toward cross-linking and chain scission rather than oxidation because of lack of locally available oxygen.

Synthetic plastic materials are generally resistant to biological attack. Diluted chemical solutions including a great variety of enzymes, acids, and bases, which readily break down natural polymers, do not affect plastics. To make plastics biodegradable they must be either filled with at least 15% of organic materials, in most cases it will be cornstarch, or the polymer chain must be composed of monomer units that themselves can be classified as nutrients for microorganisms. One such polymer, polyhydroxybutyrate-valerate, has been introduced for biodegradable films. The polymers in starch-filled plastics are initially not affected. Their breakdown follows after elution of the starch when other chemical reactions attack the much enlarged exposed polymer surface. To speed up the process, small amounts of a catalyst are often admixed into these plastic compounds.

The methods listed above for accomplishing controlled polymer degradation help to eliminate plastic litter but will destroy plastics buried in landfills only after a very long time.

Polymer degradation can also be employed to recycle some of the starting materials that have formed the polymer. Under pyrolysis conditions in which plastics are heated under exclusion of air to 900–1400°F (500–750°C), some plastics will—to a certain extent—revert to the monomer from which they were made (e.g., polystyrene, polymethacrylates, and polyesters in the presence of methanol).

Other plastics such as the polyolefins and polyvinyl chloride will form methane, ethylene, and benzene when subjected to pyrolysis conditions. In the presence of catalysts or some reactive gases the reaction can be directed so that other useful chemical intermediates are obtained at lower temperatures.

To recover the mineral fillers and reinforcing fibers, thermoset scrap or waste, including sheet molding compounds, can be subjected to pyrolysis conditions at 1400°F (750°C). The organic part, being converted to a gas, provides more than sufficient fuel to maintain the process while keeping the temperature below the melting range of the inorganic fillers.

At present the main obstacle for the exploitation of such processes lies in the lack of reliable availability of a uniform plastic waste source.

There is no question that the *recycling of plastics* is becoming of increasing importance. The burying of used plastic parts is definitely undesirable and represents a waste of a resource that has been derived from a natural supply. Extensively used beverage containers, such as the polyethylene milk jugs and the stretch-blow molded polyethyleneterephthalate soft drink bottles, are already recycled, although primarily reused for nonfood packaging applications. The multitude of compounding ingredients and mix of polymers complicate the reprocessing of the great number of plastics materials. Since for many applications it is still specified that no clean regrind of the same material can be used, applications for postconsumer recycled plastics will remain limited. Eventually even these parts, recycled one or more times, must come to the end of their usefulness, unlike metals, which are not limited in their recyclability.

The next prospect for their reapplication consists of regaining some of the chemical components by chemically reforming the polymer molecules into smaller molecular entities via cracking and distillation processes. Also, since all plastics at one stage originated from fuel materials (natural gas, crude oil, or coal) they should be redirected back to that type of material at the end of their life cycle, to regain at least their intrinsic value as fuel. For these waste-to-energy processes only a limited effort for segregation and cleaning is required since the other commingled organic materials such as paper, cardboard, rubber, and food residues also contribute to energy gains.

The fears that poisonous ingredients could be liberated into the environment should be alleviated by significant reductions in the use of toxic stabilizers and pigments, a trend that is already being pursued for many other reasons. In European countries there is a trend to reduce the total amount of chlorine and bromine in plastics, mainly to alleviate problematic combustion products. In addition, ashes and gaseous combustion products could be made subject to the separation and chemical bonding processes increasingly utilized by coal-burning power plants.

POLYMER TOXICITY

The high-molecular-weight polymers that represent the major component of common plastic materials are nonreactive in ordinary biological environments. Many of them are used for food packaging and others for medical devices. Some polymers that have been prepared under rigorously clean conditions to prevent contamination also find uses as body implants. However, all plastics are made of highly chemically reactive, low-molecular-weight compounds commonly called monomers, some of which are highly toxic. Other materials, including catalysts, emulsifiers, and solvents, are added to perform the polymerization. Supplementary materials such as stabilizers, lubricants, pigments, and fillers are introduced to the polymer during compounding. Some of these additives could be toxic and others innocuous. The remaining monomers and catalysts are now thoroughly removed from most plastics. In regard to stabilizers and other processing agents, a selection must still be made for food contact plastic materials. Even accidentally swallowed regular plastic parts which might contain toxic additives will not have a toxic effect since their migration and solubilization are very unlikely. Concern may be aroused for workers in localities in which toxic additives are incorporated into plastics as loose powders; the additives are later liberated by slowly leaching out in landfills or during incineration and the disposal of ashes.

In case of fire or when plastics are accidentally overheated during processing, toxic gases or volatile reactants may evolve. In regard to toxicity, carbon monoxide is still considered—as in other fires fueled by wood and natural or synthetic fibers—the main cause for human tragedy. Hydrogen cyanide, another very toxic gas, is formed at much lower concentrations and is also not restricted to burning plastics. In this connection it must be stated that variations in burning conditions affect combustion gas toxicity more than any other variable (e.g., acetal resins burn odorless under excess of air, forming only carbon dioxide and water, as does paraffin when burned as candles). On the other hand, under pyrolysis conditions (oxygen deficiency) toxic formaldehyde is formed. Smoke development is also of importance since it may obscure exit passages. The presence of halogen-containing compounds, especially vinyl chloride, is occasionally prohibited, to prevent potential damage. At higher temperatures these compounds can form hydrogen chloride gas, which is acrid and corrosive.

References

W.A. Combellick, Barrier polymers. In Encyclopedia of Polymer Science and Engineering, 2nd ed., H.F. Mark, N.M. Bikales, C.G. Overberger, and G. Menges, eds., Vol. 2, p. 176. John Wiley, New York, 1985.

J. Comyn and R.J. Ashley, eds. Polymer Permeability. Elsevier Applied Science, New York, 1985.

C. Hall, Polymer Materials, an Introduction for Technologists and Scientists, 2nd ed. John Wiley, New York, 1989.

L. Mascia, Thermoplastics: Materials Engineering, 2nd ed. Elsevier Applied Science, London, 1989.

G. Nelson, ed. Fire and Polymers. ACS Symposium Series 425. American Chemical Society, Washington, D.C., 1990.

S.P. Nemphos, M. Salame, and S. Steingiser, Barrier polymers. In Encyclopedia of Polymer Science and Technology, H.F. Mark and N.M. Bikales, eds., Suppl. Vol. 1, p. 65. John Wiley, New York, 1976.

W.R. Vieth, Diffusion in and through Polymers. Hanser, Munich 1991.

11

Electrical and Optical Properties
of Polymeric Materials

ELECTRICAL PROPERTIES

For engineering applications plastics share high electrical resistivity with ceramics and naturally occurring insulating materials such as mica, silk, cellulose (cotton and paper), oils, and air. These materials are contrasted to electrically conductive materials such as metals, soil, and aqueous salt solutions. Still, many plastics are formulated in such a manner that they are able to bleed off static electricity or even contribute to electrical conductivity.

Electricity can interact with insulating materials in three clearly distinguishable ways. One refers to current leakage governed by the electrical resistance of the insulator. The second involves a sudden dielectric breakdown, usually resulting in the destruction of the insulator. The third is manifested by electrical energy losses governed by the dielectric properties of the material in an alternating current circuit.

Electrical Resistivity

In general all polymeric materials must be considered good electrical insulators. Only a very few basically conductive polymers have been synthesized, with their applicability most likely restricted to semiconductors. Such commercial plastic materials consist of polymeric resins containing a variety of compounding ingredients that can affect electrical properties to a lesser or greater degree.

The electrical resistance requirement for an insulator should be set according to the maximum allowable leakage current that—for example—could flow when, in the case of a housing for an electric drill, the primary conductor insulation shorts out and the drill motor frame assumes full line voltage. The current that would leak through the molded housing—which is grasped by both hands of a grounded person—must be below the threshold for an electric shock.

Resistance values for plastics and other insulation materials are listed according to ASTM D 257 as resistivity values. Volume resistivity is determined on a cube with electrodes on opposite faces and is expressed as $\Omega \cdot m$ (area divided by height), whereas the surface resistivity is determined by two electrodes positioned on opposite edges of a square, canceling all length dimensions. Surface resistivity is expressed in Ω or Ω/square. To obtain resistance values, resistivity data must be multiplied by factors that take the particular part geometries into consideration. Since all plastics have such high resistivity values (10^8 to 10^{14} $\Omega \cdot m$), calculations based on any plastic part will produce a ridiculously low leakage current. The resistivity of even the cellulosics, at the low end in regard to resistivity, is still 1000 times higher than that of wood. However, the current that could be conducted by a plastic part showing a slight surface contamination, such as sweat or some shop soil, probably will be larger by a magnitude of 10^3 to 10^6. This dominating condition renders the small differences among various plastics practically insignificant.

Since published values can be duplicated only under conditions of utmost cleanliness and by obeying all detailed standardized test conditions, they bear little resemblance to values prevailing under ordinary working conditions. Discrepancies become especially large with moisture-sensitive plastics such as the nylons, cellulosics, and ionomers.

Even under extreme conditions where glass and porcelain have insufficient resistance, most plastics will provide excellent service. One caution must be noted: all insulators with the exception of a very few ceramics have a reduced resistance at elevated temperature. Glass insulators become conductive in the molten state.

For sensitive electronic applications such as printed circuit boards and electronic encapsulations, where high resistance values are of great importance, test screenings become necessary to obtain the most suitable plastic.

Resistivity values for various materials are listed in Table 11.1.

Electrical Conductivity

Since extensive research efforts are carried out on inherently conductive polymers, the basics in regard to the molecular requirements should briefly be noted. Polymers are selected for applications in which either their particular electrical properties—such as electrical conductivity, semiconductivity, piezoelectricity (electrets), and photoconductivity—or the convenience with which they can be processed into various shapes can be utilized.

TABLE 11.1.   Surface Resistivity of Various Materials

Material	Surface Resistivity (Ω/square)
Neat plastics	10^{12}–10^{18}
Antistatic plastics	10^8–10^{12}
Static dissipative plastics	10^3–10^8
EMI shielding plastics	10^0–10^3
Carbon products	10^{-1}–10^{-2}
Metals	10^{-5}–10^{-6}

Compounds containing only carbon single bonds (such as polyethylene or diamond) and those with isolated carbon double bonds (such as natural rubber) are good electrical insulators. When a number of carbon double bonds are alternated with a number of carbon single bonds, the more mobile electrons of the double bond become capable of traveling freely within those parts of the molecule consisting of conjugated double bonds (alternating double and single bonds).

One of the simplest organic molecules containing such conjugated double bonds is benzene. It has previously been shown (Chapter 3, p. 31) how these electrons can resonate around that ring. Since these mobile electrons cannot exit from the ring, no electrical conductivity can yet be established. However, the mobility of these electrons render this molecule capable of absorbing electromagnetic waves in the ultraviolet range efficiently whereas the more fundamental molecule of hexane remains transparent. Increasing the size of the molecule by combining several conjugated double bonded entities enhances their light absorption. The intensive colors of all organic dyes and pigments are based on this phenomenon.

If acetylene is polymerized, a structure (1) consisting of a very long stretch of conjugated double bonds is achieved, resulting in optimum resonance. The color of this polymer is black, indicating that all visible light is absorbed. Only thin films may still appear transparent or translucent as is the case for very thin, electrically conductive, metallic films. Still electrical conductivity is not quite achieved since the electrons, although mobile, remain trapped within the macromolecule. The structure of graphite was depicted in Chapter 3 (p. 31). It consists of a massive accumulation of carbon hexagon rings of a similar type as present in benzene. From the presentation of the structure of graphite it can be gathered that the resonance electrons at the edges of graphite can lead to electrical conductivity when metallic electrodes touch the surfaces of graphite.

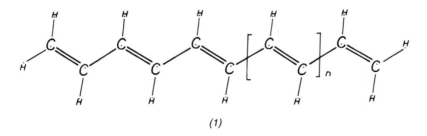

(1)

To make polyacetylene (1) or other conjugated double bonds containing polymers electrically conductive, it is necessary to incorporate metallic or metalloid ions. Their orbiting electrons or electron vacancies may facilitate interchain transfer of electrons via "hopping" of charge carriers from one macromolecule to an adjacent one. Electrical conductivity of this system is eventually secured by the transfer of electrons at the boundaries of the plastic to a metal. Other electrically conductive polymers are derived from some sulfur- or nitrogen-containing monomeric compounds such as thiophene (2a) (a five-membered ring compound with aromatic characteristics) or aniline (2b).

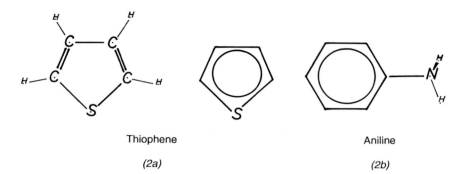

Thiophene

Aniline

(2a)

(2b)

In the case of thiophene, the polythiophenevinylene *(3)* would look similar to the long stretch of conjugated double bonds shown for polyacetylene *(1)*.

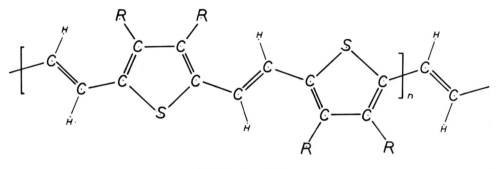

Polythiophenevinylene

(3)

Although the neat aniline polymer has a black color, at low concentrations admixed into other plastics, a translucent, faintly light green-colored plastic film or sheet can be obtained, having sufficient electrical conductivity to bleed off static electricity.

Dielectric Strength

The breakdown of plastic insulating materials because of excessive voltage gradients is important to the designer of electrical apparatus and machinery. This type of failure will in most cases determine service life expectancy. Considering the electric drill example again, the plastic drill housing will have such a low voltage gradient (only a few volts per mil) that this type of failure can never occur. On the other hand the very thin insulation (also a polymer) of the magnet wire in the motor may become exposed to 100 V/mil (4 kV/mm) at high stress points. This may lead to corona or electrical discharge currents that—although very minute—can in time chemically destroy the insulation material. Because of reduced resistance under rising temperature, this self-accelerating destruction can quickly lead to failure under overload conditions. This type of failure renders the motor inoperative but will not pose any threat to the worker.

Again related published values, generally listed under short time dielectric strength properties, are very high for all plastics and lie very close together between 400 and 1000 V/mil (16 and 40 kV/mm).

These values are not suitable for design purposes since they do not take the effects of temperature and moisture variations into account. In applications in which dielectric breakdown occurs, the cause can usually be traced to exposure of the insulating material to excessively high temperatures, either caused by high ambient temperatures, lack of adequate ventilation, or high current densities caused by mechanical or electrical overload.

The lack of adequate space for electrical equipment within a given housing frequently compels the designer to reduce its size, resulting in the need for insulating materials capable of withstanding higher working temperatures. For this important design parameter, electrical insulating materials are subdivided in a temperature classification system in which polymeric plastic and elastomeric materials are categorized from Class A, B, F, and H, based on extended experimental tests and actual work performance.

Although test values are given in volts per mil, values for different thicknesses cannot be compared since thicker films will exhibit appreciably lower dielectric strength values. For the selection of materials other tests such as the cut-through test or the determination of pinholes provide more reliable data. In the cut-through test, insulation failures are observed as crossed conductors are loaded at rising temperatures in a press until the wires short out. The quality of film or wire insulation can be determined by counting the number of pinholes per square foot (square meter) or per 1000 feet (kilometers), respectively.

Arc Resistance

Another failure mode that may occur under unusual circumstances is the degradation of the surface of an insulator by an electric arc. The arc may originate during the opening of a contact, as a result of a loose wire connection, or a reduction in the effective insulation gap caused by the accumulation of conductive dirt or moisture. Eventually material failure will occur by carbonization of the plastic surface when it is subject to the localized heat of the arc. This failure mode is easily recognizable by a lasting, black carbon track on the surface of the plastic. Although cleanliness of both the specimen and the test electrodes can affect the results, arc resistance test data can be reliably used for design purposes. Depending on the chemistry of the plastic, great differences exist among the various plastic families. There is only one plastic, polytetrafluoroethylene, that earns a near perfect score since it will not form a conductive track. An arc forming on a contaminated polytetrafluoroethylene surface will erode a small amount of the plastic, restoring a clean insulating surface.

According to ASTM D 495 the severity of the arc is greatly increased every 60 sec, therefore a material with a value of 130 sec may be a hundred times more resistant than one with a value of only 110 sec. Although it is precarious to place materials into groups, since compounding ingredients have a great influence on performance (addition of alumina trihydrate results in an exceptional improvement), a few examples should be cited. Phenol-formaldehyde and polyvinyl chloride plastic parts, though near the bottom of the list, have been extensively used for high-voltage ignition parts in automobiles in spite of frequent failures. Modified polyphenylene oxide, acrylonitrile–butadiene–styrene, ionomer plastics, and epoxy resins are positioned next. Good performance is shown by polyester and melamine-formaldehyde moldings. The best materials are represented by cellulose acetate, acetal, polyolefins, and the high-temperature polyimide plastics. Although many ceramic-type materials cannot form conductive carbon tracks, under power arc conditions the localized high heat can crack them. Applications have been cited in which plastic replacements have outperformed the ceramics.

Dielectric Loss Properties

The dielectric behavior of plastic materials is well understood and, since the *dielectric constant* (*permittivity*) of any material represents the sum of individual values for all the elements and the chemical bonds contained in them, the published values are dependable. The values for the dielectric constant of plastics at 60 Hz and room temperature range from 2 (for the highly symmetric fluorocarbon and hydrocarbon polymers) to 7 for some highly polar materials such as melamine-formaldehyde resins, polyvinylidene fluoride, and cellulose nitrate plastics. The lowest possible value is 1, which has been assigned to vacuum or air; water, at 80, is positioned at the top, overshadowed only by a few other specialty materials.

There are no good or bad dielectric constant values, since a low dielectric constant is essential for those applications in which a low capacitance is desirable (e.g., telephone cables or microwave dishes). On the other hand a high dielectric constant material is preferred for the manufacture of capacitors, since the storage capability of a capacitor increases proportionally with the magnitude of permittivity.

The relationship between dielectric constant and dissipation factor is often not well understood. Occurrences on the atomic level are therefore illustrated in Figure 11.1 and more fully explained. During *electronic polarization,* when an electric potential is laid onto the plates, the electric field, which penetrates all dielectric materials, will cause a minute shift of the electrons surrounding the positively charged atomic nuclei toward the positive electrode. Since the electrons possess practically no mass, this shift is instantaneous and hardly consumes any energy. Since the low inertia electrons can respond up to a frequency of 10^{16} Hz, only minor fluctuations are observed in permittivity values across wide temperature or alternating current frequency ranges.

To determine the dissipation factor, an alternating electrical potential must be applied. Polar substances contain molecular dipoles that consist of atoms or groups of atoms in which the centers of the positive and negative charges are spatially separated (e.g., chlorine in polyvinyl chloride). The flipping electric field twists these dipoles and moves them along the lines of the imposed chang-

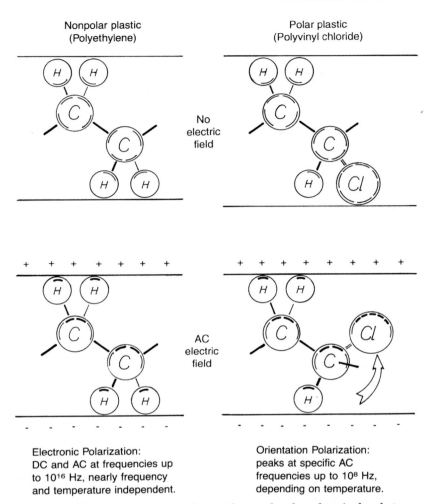

FIGURE 11.1. Dielectric responses of nonpolar and polar chemical substances. The sectors inside the circles indicate the most likely positions of outer shell electrons.

ing electrical field. In this *orientation polarization*, heavy atom nuclei must be set in motion. The amount of power consumed depends on the mass, frequency, and temperature. The displacements are time delayed as a result of the material's internal friction. The maximum amplitudes of power dissipation are obtained only when test frequency coincides with the completion of the displacements of the atoms. At higher frequencies (above about 10^8 Hz) the direction of motion is reversed prematurely prior to reaching the point of maximum displacement. At low frequencies the atoms effortlessly follow the slow pace of displacement with much time to rest at either extreme position. Therefore the energy consumed shows a maximum at one or more distinct frequencies. In extreme cases, this heat can lead to thermal destruction of the insulation material, eventually resulting in electrical failure.

The various expressions for these phenomena and the different ways in which they may be determined may also lead to confusion since the numbers, although expressing the same fact, are quite different: dielectric power factor, cosine of dielectric phase angle, sine of dielectric loss angle, dissipation factor, loss tangent, and tan δ. Some of these values also include the permittivity of that dielectric.

For certain applications dielectric heating is being utilized for heating, melting, and curing of adhesives and composite molding compounds (usually highly polar substances) with preheating of thermoset molding compounds representing the primary use.

Selection of Plastic Materials

Whenever an insulation material is selected, several possibilities for insulator breakdown must always be considered. It is not sufficient to think only about the three electrical failure modes described:

1. insufficient electrical resistance (shock hazard),
2. breakdown caused by excessive voltage gradient (corona discharge, puncture, electrical creepage and arcing), and
3. dielectric properties (leading to energy losses).

Much more attention must be paid to the expected mechanical forces, including abuse during assembly and, in particular, possible temperature excesses in both directions. Additional complications may arise when unusual environmental conditions must be met: exposure to weather and ultraviolet radiation, high humidity, or operation under water.

Since air is a fairly good insulator it is still extensively used (e.g., for high-voltage lines, in motor commutators, and in circuit breakers). In many applications plastic insulators often serve only as spacers between electricity-carrying conductors. Since contamination may pose a problem, maintenance cleaning must be planned under those conditions.

For quality assurance several different testing procedures are employed. Although high potential testing provides good certainty that the built equipment is free of defects, testing should not be overdone, since these high stresses may harm the insulation material and decrease the equipment's life expectancy.

Static Electricity

The outstanding properties of plastics materials that favor their use for electrical insulation purposes have, however, some disadvantages. Plastics surfaces will easily become electrically charged when in contact with other materials or may already be highly charged when they are removed from the mold. Static electricity attracts dirt particles to the surface of molded parts, hinders untroubled handling of thin films, and can damage sensitive electronic circuit

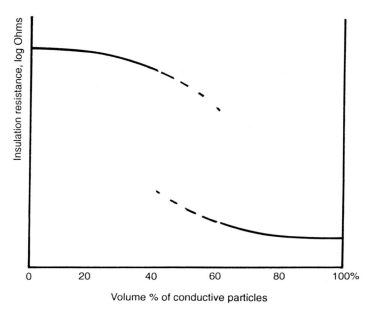

FIGURE 11.2.   Insulation resistance of metal-containing plastics.

components. Although it poses no problem to obtain a desired insulation resistance value (about 10^{10} Ω/square) under exactly controlled environmental conditions by means of the incorporation of antistatic additives, for practical purposes these treatments are often insufficient, excessive, not consistent, or not persistent enough.

In other cases even lower resistance values (about 10^3 Ω/square) are required to obtain plastic housings that can comply with electromagnetic interference and radiofrequency energy attenuation requirements. Such low values can be obtained only by incorporating good electrical conductors (e.g., conductive carbon black, metallic powders, or metal-plated fibers). The aspect ratio of fibrous fillers greatly affects electrical properties. Unfortunately, it is again not possible to reliably obtain a plastic that is a "little" conductive. As illustrated in Figure 11.2, the resistance values change very slightly when a small amount of the plastic's volume is occupied by a conductive material. Starting at the other side, the incorporation of a small amount of plastic to the conductive component will reduce conductivity just slightly. Connecting both lines in the center—as shown in most publications based on average values—cannot be justified, since materials reliably exhibiting these desirable values are difficult to obtain. Molded parts containing materials that follow the lower line must still be considered to pose electrical shock hazards, similar to painted metallic parts. Therefore in many cases other solutions for solving the problem have been found: application of electrically conductive paints to the inside or metal plating or metallic foils applied to the inner surface. Metallic or conductive screens can be combined with transparent plastics for applications in which light transparency is required.

Optical Properties

Optical properties of plastics are perceived differently depending on the particular interest of the viewer. The various subjects could include color, gloss, surface texture, and translucency, as well as special specular properties enabling them to be used for distortion-free glazing, optical lenses and mirrors, and fiber optic applications. Only a cursory account of many important aspects can be given in this chapter.

Light Transparency

Among the many structural materials only a few minerals, glasses, or plastics can form strong, clear, and colorless structural substances. In regard to *light transparency,* the graduation can extend from a colorless, transparent nearly invisible part over translucency to a complete opaqueness, including all shades of colors with their extremes to both white and black dominated tones. Since—except for a slight absorption at the blue end of the spectrum—hardly any polymer selectively absorbs light waves in the visible region (between 380 and 760 nm), colorants in the form of dyes or pigments must be added. However, clear transparent objects cannot be obtained with pigments since a certain amount of light will always be scattered by the pigment particles.

The yellowness of some plastics has been overcome in many instances by meticulous purification of starting materials and selection of specific stabilizers. Still, small amounts of blue dyes are frequently added to compensate for yellowness or its appearance after irradiation and prolonged exposure to elevated temperatures.

Surface Reflectivity

Surface reflectivity is governed by the angle of incidence and by the difference in the refractive index between the two media. The refractive indexes of some important transparent materials are given in Table 11.2.

Because of nearly the same difference in the refractive indexes (about 0.5) for all plastics and glass in comparison to air, a light ray will lose about 6% of

TABLE 11.2.   Refractive Indexes of Transparent Materials

Materials	Refractive Index, n
Air	1.000
Water	1.33
Cellulose acetate	1.48
Polymethyl methacrylate	1.49
Ordinary glass	1.52
Polystyrene and polycarbonate	1.59
Flint glass	1.65
Diamond	2.42

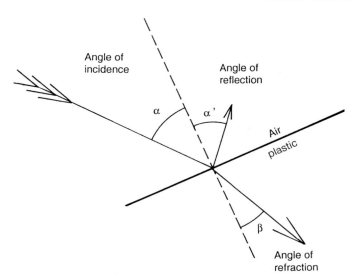

Angle of
incidence

Angle of
reflection

α   α'

Air

plastic

β

Angle of
refraction

FIGURE 11.3.   Basic optical rules.

its energy when passing head-on at a zero degree angle through any plastic or glass object as a result of light reflection. At oblique angles the amount of light reflected will increase. These losses can be lowered only by surface modifications that allow the differences in the refractive index to be gradually distributed over a certain depth. The visually perceived *gloss* of plastic parts depends not only on reflectivity but also on the smoothness of the part's surface. Therefore highly polished rolls or laminating plates are essential for obtaining glossy sheets or films. However, only by thin metallized coatings can the glossy appearance of metal parts be imitated. To obtain good durability, usually the backs of clear plastic parts are metallized.

The basic optical rules are compiled in Figure 11.3. The angle of reflection, α', is identical to the angle of incidence, α, with the angles always determined to originate from the line perpendicular to the plane. The angle of refraction, β, is smaller for materials with higher refractive indexes in a ratio depending on the difference in the indexes of both materials. The same rules also apply where the light exits the plastic material.

The incoming light will be further diminished by the presence of particles such as flattening agents, pigments, crystallites, or contaminants. Depending on the particle's nature, size, and concentration, some of the light will be absorbed but also will be scattered backward or forward, giving rise to the appearance of haze, translucency, or opaqueness. Surface texture, as already mentioned in connection with gloss, will greatly affect the quality of specular properties, since the light eventually transmitted will emerge traveling parallel or slightly divergent—in the same manner as the incoming light—only when both surfaces are absolutely smooth and parallel. Surface irregularities lead to proportionately rising light diffusion. With increasing distance between plastic sheet or film and an object, the see-through clarity rapidly di-

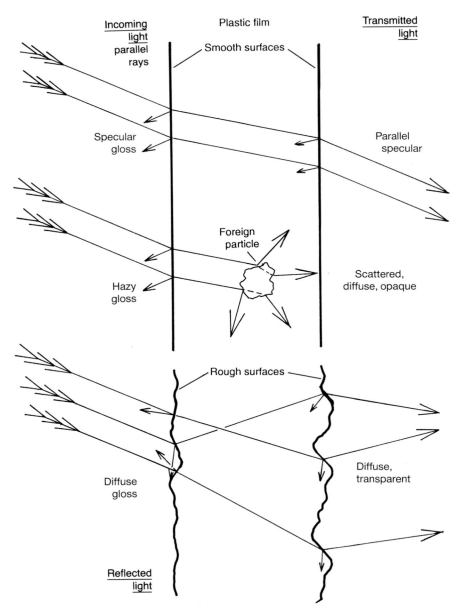

FIGURE 11.4.   Light transmission phenomena in plastics materials.

minishes, rendering many clear plastic film materials—though adequate as clear packaging materials—nonsuitable for window glazing. A few of the various possibilities are illustrated in Figure 11.4.

Originally only amorphous plastics were considered to be capable of representing transparent materials. The most important ones are the acrylics, the cellulosics, polyvinyl chloride, polycarbonate, polyesters, the family of poly-sulfones, some styrenics, and some ethylene copolymers. In addition to these, some semicrystalline plastics are now also available as transparent materials. In the case of polymethylpentene, both the amorphous and the crystalline phase have nearly the same density and refractive index; in other cases of immiscible blends, the optical properties of both phases could happen to match. Reducing the size of the crystallites below the wavelength of visible light will make some semicrystalline polymers appear clear.

Semicrystalline thermoplastics are generally translucent or opaque. However, when a certain degree of translucency is required usually clear plastics are selected and formulated with precisely proportioned small amounts of pigments or fillers. Their concentration must be adjusted in inverse proportion to the part thickness, since each particle will contribute to light scattering. One compound cannot be expected to result under all circumstances in, e.g., 20% translucency.

Color Matching

Color matching represents an important aspect in the manufacture of plastic parts. In addition to the experienced eye of a professional color matcher, there are several methods in use to physically describe any colored surface. In all these systems three numbers must be designated to define the hue—the dominant color relating to the whole visible spectrum; the chroma—indicating the intensity or purity of the color; and the value—representing the lightness or superimposed gray scale of the color, with pure black and white at opposite poles on the color sphere. Still, exact color matching is possible only when either restricted to a specifically selected lighting condition, which may simulate daylight, fluorescent, or incandescent light, or when chemically identical pigments are applied for all products to be matched in that color. The reason is that different pigments—although appearing to be identical to the human eye under one illumination—may still vary significantly in regard to their spectral absorptivity.

Surface Appearance

Surface appearance is frequently regarded as an optical quality. Both surface texturing and material coloration enable plastic parts to nearly duplicate any other material, wood, leather, textiles, stone, etc. Additional surface effects are obtainable by the incorporation of pearlescent, daylight fluorescent, and phosphorescent pigments (also called optical brighteners) or metallic flake particles, as well as by several metallizing and flock coating processes.

Plastics for Optical Components

Many more detailed optical properties such as optical dispersion, chemical homogeneity, and dimensional stability must be studied when plastics are considered for *optical components*. Further impediments such as the low surface hardness, the possibility of moisture content gradients, the high coefficient of thermal expansion, and the high molding shrinkage of plastics must be tolerated or kept under very close control. Many plastic sheets are now available as mar-resistant sheets containing various proprietary coatings. Silica and, recently, diamond coating processes for plastics are being investigated and could expand their applications into areas now still reserved for glass.

Infrared Light Absorption

Another optical characteristic of polymeric materials is their *infrared light absorption*. Depending on their chemical makeup each specific chemical group possesses characteristic absorption bands, making it possible to readily identify the chemical composition of any plastic.

Since infrared heating represents the preferred way to heat film and sheet materials for thermoforming, the selection of the proper wavelength is of primary importance in heating thin films.

Photoelastic Stress Analysis

A new field in which a peculiar optical property of plastics is being exploited is represented by the test procedure known as *photoelastic stress analysis*. Since unstrained plastic parts are optically isotropic, polarized light will not be affected when passing through them; if they are placed between a pair of crossed polarizing filters, no light will pass through. Only if the plastic part is strained, will birefringence arise. The index of refraction will become different in the two directions perpendicular to the direction of the light beam. These differences are proportional to the magnitude of strain, the thickness of the part, and a material constant. If all other parameters are kept constant, the magnitude of strain will appear as vivid colors starting with gray, white, yellow, orange, red, violet, blue, and green and followed with red and green more and more paled sequences. These observations can be applied directly to clear plastic parts to determine molded-in stresses and orientation occurring during processing or to identify dangerous stress-raisers in localized areas of the part under stress. These methods are also used to search for design improvements for metallic parts. Usually replica plastic parts are sectioned into slices of equal thickness for this analysis.

References

W.D. Callister, Jr., Materials Science and Engineering. John Wiley, New York, 1985.

M. Grayson, ed. Encyclopedia of Glass, Ceramics, Clay and Cement. John Wiley, New York, 1985.

C. Hall, Polymer Materials, an Introduction for Technologists and Scientists, 2nd ed. John Wiley, New York, 1989.

C.A. Harper, Electrical applications. In Encyclopedia of Polymer Science and Technology, H.F. Mark and N.M. Bikales, eds., Vol. 5, pp. 482–528. John Wiley, New York, 1976.

K. Mathes, Electrical properties. In Encyclopedia of Polymer Science and Technology, H.F. Mark and N.M. Bikales, eds., Vol. 5, pp. 528–628. John Wiley, New York, 1976.

R. Zund, Optical properties. In Encyclopedia of Polymer Science and Technology, H.F. Mark and N.M. Bikales, eds., Vol. 9, pp. 525–623. John Wiley, New York, 1976.

12

Compounding of Polymeric Materials

Only in exceptional cases will commercial plastics be absolutely pure, that is, not contain any compounds other than those indicated on the label. This should not be cause for alarm. All technical substances contain a certain amount of impurities. Sometimes these small amounts of substances will consist of residues not completely removed after completion of the polymerization reaction. Some of them have been introduced as polymerization aids, which are often required for a smooth conversion. Others may be solvents, dispersing agents, emulsifiers, salts, pH adjusting agents, etc. Only rarely will these small amounts interfere with further processing steps or with the plastic material's service. If potential difficulties are feared, higher purity plastics must be demanded. Examples include the plastics used for optical fibers, compact discs, fine electronics, prosthetic devices, and medical implants.

Just a few examples should be cited here to clarify this point. Traces of chlorine, originating from the starting material, may be present in epoxy resins. Small amounts of monomer may still be contained in polymers, however, vinyl chloride is now thoroughly removed to prevent exposure to this potential carcinogenic gas. Esterification catalysts may remain in thermoplastic polyesters, which might later catalyze transesterification in some plastic blends. Benzoic acid derivatives remain in resins polymerized with benzoylperoxide. Traces of acid stemming from the curing catalyst may affect properties of phenol-formaldehyde resin adhesives. This can be eliminated by changing to resorcinol-formaldehyde resins, which do not require acids for curing. Acetic acid will become liberated from many RTV silicone rubbers during moisture cure.

The presence of special compounding ingredients is generally noted by the plastics materials suppliers. However, since plastics are sold according to their intended processing and their application decreed properties, the chemical composition of compounding ingredients is very seldom mentioned, only their salient properties, such as "weatherable," "plating grade," or "flame retardant." Attention must sometimes be paid to negative descriptions: "Does not contain chlorine" can mean that the plastic does not contain any halogen atoms or that it contains bromine. The formulating ingredients are considered proprietary information and are in most cases not declared. Frequently the chemical nature of an additive is cited by the manufacturer only after it has fallen into disfavor and has been replaced by an improved compound (e.g., "does not contain heavy metal compounds"). Its new formulation, in turn, is not divulged and its chemistry is again listed as proprietary. When qualification tests for a certain plastic compound have been accepted, these tests should be repeated when an "identical" compound from a competitor is chosen as an alternate. In particular, the ingredients in flame-retardant plastics may pose constraints in regard to processing and also in mechanical properties.

In recent years the selection of compounding ingredients has increased in importance because of their potential toxicity. The complexity of this subject can be demonstrated by the quite different manifestations of this problem. Controlling the exposure of the worker to dusty, harmful additives during the manufacture of a compounded plastic is simple. Many of these additives are now also available as dust-free powders. The leaching of toxic additives is of significance only for food packaging and medical applications. Within the past few years the release of toxic additives after a part is discarded has become of concern. Toxic chemicals may leach out—in time—when a part is discarded in a landfill or may appear as an air pollutant or in the ashes when parts are incinerated under improper conditions.

Other problems can develop with additives as a result of undesirable and sometimes unexpected behaviors. Sometimes their limited compatibility can cause discolorations, chalking, blooming, and migration to the surface. Some lubricants may exude and form a sticky surface and some pigments may plate-out at the mold surface. Other ingredients, such as stabilizers and plasticizers, may not be permanent enough, and still others may infringe upon the odor.

Also the purpose for the addition of compounding ingredients will vary. The objective of many additives is to facilitate processing of the plastic by the molder or by the converter of film or sheet products. In a few cases the additive, an antistat, helps eliminate dust attraction on the surface of molded parts while on the store shelf. Most additives, however, serve to provide the desirable surface appearance and to maintain the part's durability throughout its expected service life.

Descriptions of compounding ingredients, especially impact modifiers and fiber reinforcements, can be found in other chapters of this book.

PLASTICIZERS

In most applications in which plasticizers are incorporated into plastics, their purpose is to convert an otherwise hard and rigid plastic to a flexible or semiflexible tough part. In some instances the same or similar conversions are obtainable by copolymerizing certain flexibility imparting monomers or by blending or grafting elastomers with those rigid plastics. Some of them, such as a number of acrylonitrile-butadiene or ethylene-vinyl acetate copolymers, are specifically made for use as polymeric-type plasticizers.

The incorporation of a plasticizer, which in most cases is a low viscous liquid, is easier to accomplish and much more flexible than formulating copolymers.

When selecting liquid plasticizers that possess many of the typical characteristics of solvents, their chemistry must be taken into account to achieve compatibility with the polymer. Only a few polymers are able to absorb large amounts of plasticizers and still retain sufficient strength. In such cases, strong, polar bonds, extending from the polymer chain, hold adjacent chains together at certain intervals, whereas bulk parts of the originally rigid chain will be rendered quite mobile. In plasticized polyvinyl chloride, the most important representative, the weight ratios of polymer to plasticizer for the flexible compounds are established at about 60 to 40 and at about 75 to 25 for the semiflexible compounds.

The low-molecular-weight plasticizers show the best plasticizing performance but have the disadvantage of being more prone to migration when in contact with other plastics and of being somewhat more volatile. The plasticizing power is proportional to the viscosity of the plasticizer or mixture of plasticizers, with the lowest viscosity resulting in the best flexibility. Evaporation loss of plasticizers render plastics less flexible in time. The plasticizer vapors may cause windshield fogging in automobiles because they condense on the interior glass surfaces.

Although most plasticizers are true plasticizers, meaning that they are completely soluble in the plastic, many other less compatible substances are retained only in a dispersed (not dissolved) state in the plastic, although more integrated than some of the elastomeric impact modifiers. These plasticizers are generally of higher molecular weight. Others may occasionally work themselves slowly to the outside, imparting an oily surface to the part.

The effectiveness of any plasticizer in regard to imparted mechanical properties is very much temperature dependent. That is why most plastics are not compounded with plasticizers. At elevated temperature they would become too soft and at the low temperature range they would be too hard and brittle. The cellulosics and polyvinyl chloride represent notable exceptions. Their plasticized formulations will perform well within the important temperature range of $-40°F$ to $160°F$ ($-40°C$ to $70°C$).

This great temperature dependence is exploited by the processors of many plastics. When only small amounts of a plasticizer are compounded into the plastic (see also p. 232), or when very high viscous or solid plasticizers are added, or when polymers containing a small percentage of lower molecular weight fractions are processed, the melt viscosity will be markedly lowered. This will greatly facilitate melt processing without affecting any of the mechanical properties at the much lower room temperature. This tampering, however, can easily be discovered by conducting acceptance tests not only at room temperature but also at elevated temperatures, at least at the upper limit of the expected working temperature.

Plasticizers vary widely in chemical composition, although the most frequently used esters represent the most suitable class of primary plasticizers. The phthalate esters have the best solvating power, whereas the phosphate esters assist in the preparation of flame-retardant compositions. Other ester plasticizers are the entirely aliphatic adipic acid esters (adipates and also the azelates and sebacates), which exhibit exceptionally good low-temperature performance (but lower permanence), and the polymeric plasticizers consisting of diols and diacids, which top other plasticizers by their extraction and migration resistance and permanence. The higher cost terephthalate and trimellitate esters take an intermediary position.

Another group of plasticizers, generally termed secondary plasticizers, lack compatibility and are therefore jointly used with good plasticizers. They may consist of high-boiling aliphatic or aromatic (some also chlorinated) hydrocarbons of an oil or wax character. Three special plasticizers are the epoxidized oil plasticizers, which also act as heat stabilizers for vinyls, the old established toluenesulfoneamides, and the citric acid esters, which are chosen when food contact becomes a possibility.

PLASTICS ADDITIVES

A great number of different compounds are employed as additives for plastics. In many cases they are designated according to their purpose, such as stabilizers, lubricants, or blowing agents. The number of compounds is so vast because some have only a very narrow range of application and are limited to only a few types of plastics, a few may have accompanying weaknesses, and still others have multifunctional serviceability. For the latter reason their chemistry should be dealt with collectively.

To emphasize the multiplicity of the duties of additives, they should first be listed. A variety of factors influence the appearance of the part, including the various pigments and dyes with all the variations of iridescent, fluorescent, pearlescent, metallic, and optical brightener characteristics. Some formulations require dulling agents and others require gloss-enhancing additives. Many additives come in powders, color concentrates, or with other additives combined as master batches to improve dispersibility. Pigments and dyes are further subdivided into groups depending on their heat stability, light fastness, weatherability, acid resistance, and tinting strength and on whether they are nonmigratory, nonbleeding, and easily dispersible. Additional features could further be included to ensure lack of mold plate-out or surface bloom and also to protect the part's surface by the addition of antistats or antifogging agents. The latter are mainly of importance for maintaining clarity of plastic packaging by preventing condensation of water droplets. Other additives provide mold release and antiblocking and slip properties, and prevent mildew formation (antimicrobials).

Processing aids are also manifold and are used to reduce melt viscosity and increase hot strength so to eliminate melt fracture and tearing in extrusion processes. In some instances an addition of only 0.1% will markedly reduce extruder torque requirements and increase output. In plastisol formulations viscosity control is also of importance (thixotropic or nonthixotropic). A great number of mold release agents are required depending on whether they are applied to the mold surfaces (metal or epoxy plastic molds) as water- or solvent-based sprays or whether they are formulated with the resin. Some must convey antiblocking properties; others need correction of the coefficient of friction by adding either slip or antislip agents.

Stabilizers are quite universally applied additives with heat and oxidation stabilization (antioxidants) most commonly used, followed by light stabilizers, and UV absorbents. However, in a few applications just the opposite effect is required. For the production of plastics that should easily photo- or biodegrade when discarded, additives must be incorporated to achieve light (UV) or biological deterioration of the plastic within a short time.

Some additives serve very specific purposes, such as reinforcing fibers for composites, blowing agents for the production of foams, and flame-retardant and smoke-suppressant additives.

All these additives differ in regard to their specific compatibility with the plastic, their toxicity (with special grades for food and medical applications), their contribution to pollution, and their effect on product clarity and stain resistance. Their incorporation into the plastic may depend on whether they are available as free-flowing and low-dust powders or as liquids or in master batch form.

Fillers

As the name implies, fillers, in most applications are used to lower the cost of the plastic. The modulus of elasticity of plastics is increased when fillers are used, however, tensile and impact properties are in most cases reduced. Their presence is indispensable only in highly cross-linked thermoset resins in which they are used to control cure shrinkage. Lower particle sizes (submicron grades) are generally more beneficial in improving mechanical and optical (opacity) properties. The loading of fillers in plastics is dependent on the shape and the size of the filler particles.

When compounding thermoplastics, the particles will have to be surface treated with coupling agents. The titanates are preferred when compounding polyolefins and vinyls. Depending on the type of the chemically reactive group attached to the silanes (amino, vinyl or mercapto), they can be suited for use in a variety of plastics.

The chemistry of any coupling agent is characterized by two types of groups, one capable of forming a chemical bond with the surface of the filler or reinforcing agent and the other capable of intermingling with the chain links of the polymer. It should be noted that any material containing ionic bonds, which includes all inorganic fillers, possesses an electrically charged and reactive surface. Only the chemically neutral thermoplastic polymers require the addition of some ionic groups or a coupling agent. Since the cure of most thermoset polymers involves a chemical reaction, an adequate bond with fillers is invariably obtained, unless the particles are coated with oils.

The surface coating applied to the filler greatly influences the oil/plasticizer absorption values and also improves the quality of the bond between filler and polymer. This is of great importance in negating the undesirable influence of fillers on some of the mechanical properties of plastics. As with pigments in paints, a low oil absorption number allows higher loadings of fillers, but increased melt viscosity will generally limit their concentrations to moderate levels.

Calcium carbonate, as one of the more important mineral fillers, is represented either as natural limestone (calcite), which has been wet ground (0.5–20 μm) or dry ground (20–44 μm), or as chemically precipitated calcium carbonate of much lower particle size (0.05–10 μm). Although calcium carbonate is nearly nonabrasive, it is sensitive to acids and may have a gray or brown color because of its iron impurity. Calcium carbonate of narrow particle size (0.1–0.25 μm) has the best hiding power and reduces the need for the expensive rutile titanium dioxide pigment.

Silica fillers are also low cost but are quite abrasive. Both amorphous and crystalline silica is ground to 2–15 μm size. Much larger particle size silicas (sand) or glass beads are used as fillers for casting resins.

Several *silicates* are commonly used. Kaolin clay is mostly applied as water-washed Georgia clay with a particle size of 0.5–5 μm. It is also available as calcined hard clay. Talc particles have a laminar structure and wollastonite an acicular structure that have found some applications as substitutes for asbestos.

Hollow glass spheres with a particle size of 10–250 μm represent a special filler for the production of lightweight objects and syntactic foam structures. Although these compounds cannot be subjected to high shear processing, hydrostatic pressures up to 20,000 psi (150 MPa) are tolerated.

Barite, a barium sulfate mineral, has a very high specific gravity and good hiding power. It is also capable of replacing titanium dioxide in some applications. It is sold as dry ground (12 μm), wet ground (3 μm), and also as a precipitated filler having a still lower particle size.

Organic fillers are primarily wood or nut shell flour, cellulose fibers, textile flock for reinforcements, and nylon or polytetrafluoroethylene powders for low surface friction compounds.

To raise thermal and especially electrical conductivity, a very high loading of *conductive fillers* is required. As can be gathered from Table 12.1, thermal conductivity cannot be improved even 10-fold by a very high percentage of a conductive metallic filler.

To obtain good electrical conductivity it is necessary not only to provide high loadings of good electrical conductors but also to ensure that their surface remains oxide free. Metal powders—unless silver is chosen—should be silver or gold plated prior to their incorporation. Metallic fibers (usually nickel) will

TABLE 12.1. Thermal Conductivity of Plastic and Metallic Mold Materials

Material	Thermal Conductivity	
	Btu · ft/hr · ft² · °F	W/m · K
Epoxy resin, no filler	0.13	0.22
Epoxy resin, 91 wt% Cu (60 vol%)	0.94	1.63
Steel	37	64
Aluminum	135	230
Copper	225	390

TABLE 12.2.   Electrical Conductivity of Some
Filled Plastics and Metals

Material	Electrical Volume Resistivity	
	$\Omega \cdot cm$	$\Omega \cdot m$
Epoxy, carbon filled	10^4	10^2
Epoxy, silver filler	10^{-2}	10^{-4}
Graphite	10^{-2}	10^{-4}
Mercury	10^{-4}	10^{-6}
Aluminum, copper and silver	10^{-6}	10^{-8}

also ensure continuous, good electrical contact between particles. Electrical volume resistivity values are given for selected materials in Table 12.2.

Pigments

Inorganic pigments are preferred for coloring plastics. They are not only more thermally stable, easier to disperse, and have higher hiding power than most organic pigments but are also lower in cost. The best yellow, orange, and red pigments, unfortunately, contain the heavy metals lead and cadmium as chromates and molybdates, which are now being replaced by organic pigments of considerably lower toxicity. Some of them may be only marginal in regard to heat stability, light fastness, and resistance to migration. To make pigments more adaptable for incorporation into the plastic, they became available as nondusting powders and as liquid or solid concentrates depending on the intended blending process (see also optical properties).

Stabilizers

The addition of a stabilizer is not necessarily the most effective way of obtaining heat stability. In many cases the exhaustive removal of polymerization catalysts or other impurities will be of equal or dominant importance.

As stated, it is often difficult to differentiate among heat, oxidation, and light stabilization. Furthermore these three might be of varying importance for the user of plastics parts. Heat stabilization is also a necessity for the processor. First the various kinds of stabilizers used should be discussed.

1. Hindered phenol antioxidants function mainly by trapping free polymer radicals that might have been created under any of those destructive influences. They consist of phenols in which the most chemically reactive sites have been blocked by alkyl or other nonreactive groups. Their simplest derivative is the butylated hydroxytoluene (2,6-di-*tert*-butyl-p-cresol) *(1)*.

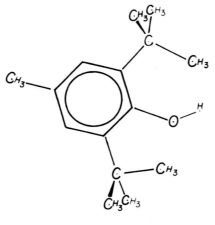

2,6-di-*tert*-butyl-*p*-cresol

(1)

2. Thioethers or sulfides can react with peroxides that have been attached to polymeric chains. This will prevent the breakup of the polymer chain. They are represented by dithiocarbonates, dithiophosphites, and organotin mercaptides (e.g., dibutyltin diisooctyl-thioglycolate) *(2)*.

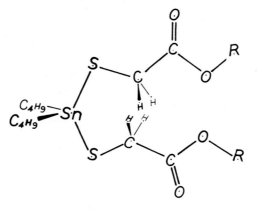

Dibutyltin Diisooctyl-thioglycolate

(2)

3. Tertiary phosphites prevent discoloration from occurring when polymer molecules start breaking down. They are exemplified by a large number of trialkylphenyl phosphites *(3)*.

4. Since metal ion impurities often initiate thermal decomposition, chelates (compounds that can tightly bond these ions) have found applications. The 2,2'-thiobis(4-*tert*-octylphenolato) nickel chelate is an example *(4)*.

5. UV absorbers will render the highly active radiation immune by converting it into innocuous heat. These compounds contain conjugated double

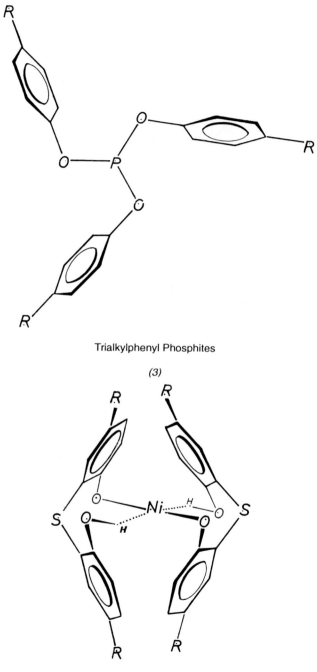

Trialkylphenyl Phosphites

(3)

2,2'-thiobis (4-*tert*-octylphenolato)
Nickel Chelate

(4)

bonds and/or hydrogen bonds that can elicit resonance within the molecule (5–8).

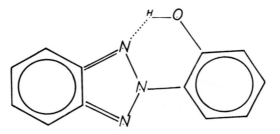

2(2'-Hydroxyphenyl)-benzotriazole

(5)

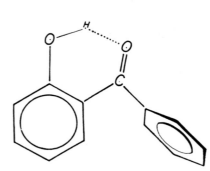

o-Hydroxybenzophenone

(6)

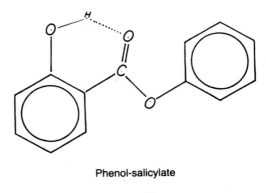

Phenol-salicylate

(7)

6. The hindered amine light stabilizers (HALS) are frequently listed separately. Since they represent rather volatile and extractable organic molecules, processes to chemically bind them to the polymer molecules are being developed. The exceptional efficiency of hindered amine light stabilizers can be attributed to their capability to trap polymer-derived radi-

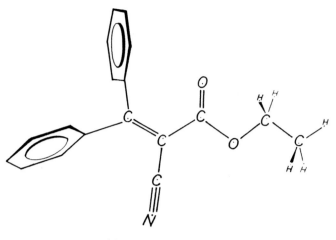

Substituted Acrylonitrile

(8)

cals by a regenerative rather than sacrificial process. These compounds, mainly based on piperidine *(9a)* or piperazine *(9b)*, are valued as light and radiation stabilizers and antioxidants especially for polyolefins.

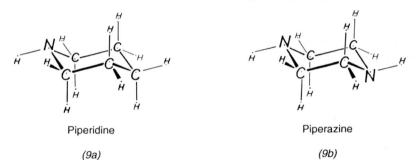

Piperidine   Piperazine

(9a)   *(9b)*

Still, one should remember that no additive can duplicate the efficiency of carbon black when it comes to the selection of a light stabilizer. In addition, it can stand in for part of an antioxidant and antiozonant. It is also the additive that is least likely to be extracted or used up in time.

Since polyvinyl chloride will readily split off some hydrochloric acid when exposed to high temperatures or high intensity light, special kinds of stabilizers have evolved for it and other halogen-containing polymers. Although barium–cadmium stabilizers provide the best performance, their toxicity has resulted in efforts to replace barium by chemically related strontium and calcium compounds and cadmium by the related zinc compounds. Tin stabilizers similar to the ones already referred to and others such as dibutyltin dilaurate *(10a)* (one of the carboxylates) share applications where clarity is of importance.

Since these metal derivatives can contribute to the deterioration of electrical properties, lead-containing stabilizers dominate where polyvinyl chloride is

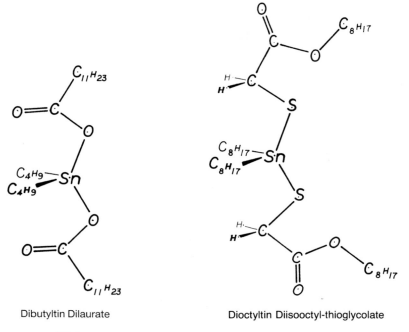

Dibutyltin Dilaurate

(10a)

Dioctyltin Diisooctyl-thioglycolate

(10b)

used in electrical insulation materials. Both the lead oxides and salts and their conversion product (lead chloride) will remain insoluble and will not contribute ionizable compounds to the plastic, thus retaining the excellent electrical properties for an extended period of time. Some of these important stabilizers are:

Basic lead carbonate	$2PbCO_3 \cdot Pb(OH)_2$
Tribasic lead sulfate	$3PbO \cdot PbSO_4 \cdot H_2O$
Dibasic lead phthalate	$2PbO \cdot Pb(OOC)_2C_6H_4 \cdot {}^{1/2}H_2O$
Dibasic lead phosphite	$2PbO \cdot PbHPO_3 \cdot {}^{1/2}H_2O.$

All lead stabilizers contribute to the opaqueness of the plastic formulations and therefore will not be applicable for plastic compounds of high clarity. In spite of much searching, no nontoxic replacements have been found.

Lubricants, Mold Release, and Antistatic Agents

These three additives also act interchangeably in several cases. Lead stearate and dibasic lead stearate work both as lubricant and stabilizer. Stearic acid, however, acts only as a lubricant, similar to microcrystalline waxes with a relatively high melting point of 165°F (75°C) and low-molecular-weight polyethylenes. For polyvinyl chloride, the ethylene-bistearamide represents an important lubricant.

Sometimes mold releases such as carnauba wax, zinc stearate, and silicone oils are incorporated directly into the plastic. Otherwise, the same but also different substances, such as polyvinyl alcohol (only in the form of solutions), can be applied to the mold surface as liquids, powders, or aerosols. For flat, molded laminates, slip sheets consisting of cellophane can be used. If decorating (painting) or bonding of the plastic part is to follow, the mold release agent must either be removed or the one being selected must not interfere with obtaining a good bond.

To achieve a permanently antistatic plastic surface, a hygroscopic substance is usually incorporated into the compound. Very small amounts of ethoxylated aliphatic amines and amides, quaternary ammonium salts, glycol esters, or nonionic surfactants are usually chosen. Their function is to attract moisture to the surface to bleed off the static charge. Other antistat additives lower the electricity resistivity of the bulk plastic. Since the amount of electricity is rather low, the surface resistivity must in many cases be reduced just marginally (10^9 Ω/square). The effectiveness of these agents is generally determined by the time required (seconds) to dissipate a static electric charge.

If static electricity becomes a problem only during processing of films at high speeds, high-voltage or radioactive static eliminators that ionize the ambient air will prove expedient.

Chemical Blowing Agents

Although in nearly 90% of cases physical blowing agents are applied for the production of cellular plastics, chemical blowing agents have preserved their niche where cellular plastics parts are produced using conventional plastics machinery. Gases such as carbon dioxide, nitrogen, and air or low boiling liquids such as pentane or still some chlorofluorocarbon liquids (now being temporarily replaced by hydrofluorocarbons or methylene chloride) are representatives of such physical blowing agents.

Most exothermic chemical blowing agents represent storage-stable organic compounds that can be compounded into plastics much like any conventional additive. Others represent inorganic compounds such as bicarbonates or borohydride. Each compound has a narrow activation temperature range at which it will chemically decompose, forming gases, primarily nitrogen, carbon dioxide, and hydrogen as well as ammonia and water vapor. It is necessary to ensure that the heat generated during decomposition and the chemicals formed will not affect the integrity of the polymer. To regulate the pore size and structure, other additives must be adjoined. Those could be activators, surfactants, or nucleating agents. The organic blowing agents carry two or more nitrogen atoms in a row or a ring structure favoring the formation of molecular nitrogen on decomposition. Only the formulas for these active sites are given here (11a–d):

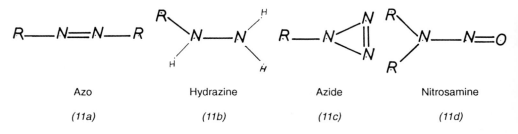

Azo	Hydrazine	Azide	Nitrosamine
(11a)	(11b)	(11c)	(11d)

The performance of blowing agents is generally characterized by their temperature range and their gas yields, which are mostly situated between 100 and 250 cm³/g.

Endothermic foaming agents, which generally consist of bicarbonates and organic acids, provide several advantages: their reaction consumes heat since they represent endothermic mixtures, they do not involve any materials not conforming to Food and Drug Administration requirements, and they simultaneously act as nucleating agents, resulting in fine pore size foams.

Flame Retardants and Smoke Suppressants

Here again the active agents are manufactured from a great number of organic and inorganic chemicals, which in many cases are combined to take advantage of potential synergistic effects. The organic compounds contain either one of the halogen atoms chlorine and bromine or phosphorus. Because of the high atomic weight of bromine, 80, versus 35.5 for chlorine and 31 for phosphorus, a smaller number of bromine atoms in a molecule or smaller concentration of the agent are required. The aliphatic hydrocarbons are mostly represented among the chlorinated compounds such as chlorinated paraffin, chlorinated wax, and the chlorinated cycloaliphatic derivatives. The brominated compounds prevail in aromatics, some of which also carry reactive groups so that they can easily be incorporated into the polymer chain. Representative compounds are octa- and decabromodiphenyl oxide and tetrabromophthalic anhydride or tetrabromo bisphenol A. Among the phosphorus-containing flame retardants the phosphate plasticizers such as the tris-arylphosphates and tris-chloroalkyl phosphates, the phosphine oxide and the cyclic phosphonate esters should be mentioned. However, elemental coated red amorphous phosphorus is also used.

The most established inorganic flame retardant is antimony trioxide, which is always used with chlorine-containing compounds. Sodium antimonate must be used when the alkalinity of the trioxide may cause polymer breakdown. Among the other inorganics, alumina trihydrate also provides good electrical properties but must be added in higher percentages. Zinc borate, molybdenum trioxide or related compounds, and ferrocene are also good smoke suppressants. The release of moisture at high temperatures represents the working principle in alumina trihydrate, zinc borate, and magnesium hydroxide, but also demands more care during processing to prevent untimely decomposition.

Most of the inorganics will promote char formation, which retards combustion because of the establishment of a thermal insulating layer.

REFERENCES

R. Gaechter and H. Mueller, Plastics Additives Handbook. Hanser, Munich, 1990.

R.B. Seymour. Additives for Plastics. Academic Press, New York, 1978.

H.R. Simonds and J.M. Church, The Encyclopedia of Basic Materials for Plastics. Reinhold, New York, 1967.

J. Stepek and H. Daoust, Additives for Plastics. Springer-Verlag, New York, 1983.

13

Composites Utilizing Polymeric Matrices

The term composites is broadly defined as any substance consisting of more than one distinct, integral component, although most engineering professionals reserve this term for oriented fiber-reinforced materials not necessarily of polymeric origin. This chapter on composites covers all plastics materials that consist of more than one immiscible material and also includes materials in which the dispersed phase does not represent a fibrous material.

Single-phase copolymers, such as vinyl chloride–vinyl acetate, are definitely not composites. All their properties are placed proportionately in accordance to the mass fractions of the constituents involved. On the other hand, copolymer constituents with limited solubility, miscibility, or compatibility consist of two or more locally segregated components. In many cases one will be the continuous phase or matrix, which contributes to one set of properties, and the other will be the phase or phases that are dispersed in the matrix, which contribute to a different set of properties. In rare cases both phases must be considered continuous. Those materials are called interpenetrating network plastics and were covered in Chapter 5. A fine-tuning of material properties can be obtained by various combinations of the different constituents as manifested by the great number of acrylonitrile–butadiene–styrene plastics.

Books on composites cannot claim to contain all information that has been obtained on this class of materials and at this time only a fraction of all possible, potentially useful combinations have been investigated. For this reason the description of composites in this book will be limited to defining the principle of composite materials, the benefits obtained when incorporating an additional phase into a plastic material, and some of the general requirements in regard to the properties both phases must present.

This chapter also briefly covers structural and insulating foams since they represent materials in which plastic matrices are combined with air or gases as a dispersed phase.

Since this book focuses on structural polymeric materials, some very important biogenic composites will also be described in detail. Although the same rules for obtaining the most favorable mechanical properties apply to all these different classes of materials, the unique possibilities as exhibited in nature occupy a prominent standing. How the monomeric building blocks are positioned at the precisely desired location during polymerization is still not known to scientists and this kind of structuring has still evaded duplication by chemists.

The intermingling of components in concrete relates well to the plastics materials discussed in this chapter, and concrete is one of the composite materials most copiously used in construction. Although it is not a plastic material, organic polymeric materials are increasingly used in its formulation both as binder additives to speed up cure and improve strength and as fibers to reduce crack propagation.

On the addition of water portland cement powder forms a hard, crystalline matrix that by itself will not provide a strong usable material because of its excessive shrinkage in volume. During the crystallization process too many cracks will form. When this matrix phase is extended up to 10-fold through the incorporation of hard and strong nonshrinking substances (sand and gravel) as the dispersed phase, crack formation can be minimized, resulting in a strong and enduring building material. Further significant improvements, especially of tensile properties, can be achieved by an additional phase consisting of steel reinforcing rods placed in the direction of highest tensile forces. The advantage of citing concrete as an example lies in the easily visible distribution of its components. To obtain similar looking illustrations of plastics composites, pictures must be taken through a microscope or electron microscope.

The oldest thermoset plastics were based on formaldehyde condensation resins (phenolics or urea resins), which all suffered from excessive shrinkage, the same weakness exhibited during cure by portland cement. The excessive shrinkage in this case is caused by the evaporation of the formed water and by further approachment of the molecule's sections during cure. Therefore, such neat polycondensation resins are unsuitable for molded parts.

At first this statement may not make sense, as thermoplastics exhibit large shrinkages in volume during both polymerization and processing. The specific gravity of polystyrene is 1.05 g/cm^3 and that of the styrene monomer only 0.907 g/cm^3. The coefficient of thermal expansion of polystyrene is $0.3-0.4 \times 10^{-4}$ in/in·°F ($0.5-0.8 \times 10^{-4}$ cm/cm·K) and that of low-density polyethylene $0.6-1.1 \times 10^{-4}$ in/in·°F ($1-2 \times 10^{-4}$ cm/cm·K). The significant differences are that with thermoplastics, the volume shrinkages occur when the growing polymer is either finely dispersed or in a liquid though highly viscous state or when the forces occurring during thermal expansion (or contraction) are uniform and can easily be absorbed by the loose arrangement between adjacent polymer chains. Even in thermoset resins, regardless of which polycondensation reaction is employed, the bulk of shrinkage takes place in the chemical reaction vessel, thus reducing material shrinkage during molding. The concurrent buildup of molecular weight is there fostered to the greatest extent possible, as long as sufficient plasticity is retained so that the resulting material remains moldable under heat and pressure. The detrimental shrinkage in thermoset resins occurs only during the final formation of the three-dimensional network comprised of tightly intermeshing covalent chemical bonds and the accompanying loss of volatile condensation reaction products.

The less rigid thermoset resins do not pose these drastic conditions since there the chemical cross-links only sparsely interconnect the generally chain-like structures. This group includes those plastics that derive from thermoplastics through irradiation or chemical cross-linking. Others represent elastomers prepared by vulcanization, or thermosets that utilize prepolymers for the formation of polyurethanes and other similar thermoset plastics.

The chemical processing of starting materials involved in the manufacture of polymeric resins has always added significantly to the cost of these materials. Therefore these resins are often filled or extended with cheap inert materials. In the case of thermoplastics, the loss in pliability (tensile strength and elongation) and optical clarity and the increased difficulty in processing have long prevented and still restrict the use of fillers. Highly filled thermoplastics (65 wt%) are now extolled for imitating ceramics both in looks and feel. Their high specific gravity of 2.4 g/cm^3 is comparable to that of glass, ceramics, and aluminum.

However, in the case of thermosetting resins, the inclusion of fillers has improved their desirable properties including compressive strength, hardness, durability, abrasion resistance, and high-temperature resistance. The greatly reduced flow of these formulations did not pose a problem since processing was and still is dominated by high-pressure compression molding, requiring only minimal material flow.

An additional difference in regard to the inert fillers used for plastics compounding must be mentioned. In the case of thermoplastics, the inert fillers were just occupying space without contributing to the formation of a mechanically coherent system. This has been changed by the availability of proprietary, specially surface-treated fillers and coupling agents that are simultaneously compounded into thermoplastic resins (see Chapter 12, p. 269). The effectiveness of these measures has been evidenced by improved mechanical properties, reduced water absorption, and diminished loss of properties experienced during extended boiling in water. This latter harsh procedure is widely used to test the stability of composites.

In the case of thermoset plastics, the final chemical reaction takes place in the presence of the fillers. Since even the most inert fillers display a chemically unbalanced surface layer, a fair bond has always been obtained, even in cases in which no true chemical bonds are established. This has been proven by the dominance of cohesive rather than adhesive failures. When compared with thermoplastics, the narrower gap between thermosets and fillers in regard to those constituents' modulus of elasticity and their coefficient of thermal expansion is probably a contributing factor.

For this reason most fillers, regardless of their shape, must be considered reinforcements, benefiting not only the compressive and flexural properties but in many instances also the tensile and impact properties of thermoset compounds. The peak performance is usually reached at quite high volume percentages of fillers. The rubber tires presently in use confirm that fine powders of no inherent strength can have very beneficial effects. The tire's excellent wear and endurance properties cannot be obtained from elastomers unless the generally high charges of carbon black fillers are present.

The changes in volume shrinkage resulting from filler loading of the formaldehyde condensation resins should be further scrutinized. Although shrinkage of the matrix phase (the thermosetting resin) will not be altered, the developing forces are considerably reduced because of the short distances in the matrix phase extending just between adjacent filler particles the dimensions of which are not changing. When additional external forces are imposed on unfilled, brittle resins, a crack—originating at the location of highest stress—would easily propagate through the entire piece. With abundant filler particles present, crack growth or crack propagation will be arrested at each filler location. For ultimate failure to occur, a new crack must be reinitiated repeatedly.

Particulate fillers are not necessarily limited to rigid, hard, high modulus substances. Toughened thermoplastics are based on the incorporation of low modulus, elastomeric filler particles compounded or chemically bonded into a thermoplastic matrix. High-impact polystyrene (or rubber-modified polystyrene) was initially formulated to reduce the susceptibility of polystyrene to brittle cracking without causing a significant adverse effect on its high modulus of elasticity and its fair thermal properties. These successes culminated in the synthesis of the acrylonitrile–butadiene–styrene copolymers, which have achieved a status of their own and now compete as engineering plastics. Among other thermoplastics available in toughened versions are polyvinyl chloride, methyl methacrylate, and nylon.

The toughening capacity is founded on the formation of crazes in the plastic matrix and the elastic extension of the elastomeric component. The effect of geometry, as described in Chapter 7, should be recalled. The small spherical rubber particles are capable of supporting an appreciable fraction of the total load on the composite as a result of their relatively high bulk modulus of elasticity in that chubby shape. Unlike a slender rubber band or a thin rubber balloon, which are highly extensible, a large caoutchouc bale of nearly cubical dimensions represents a hard, tough, hardly extensible material. The initial modulus of elasticity of elastomers is quite high between 0 and 3% elongation, as seen in Figure 7.1 (p. 188). On the other hand, stone-hard mineral particles will not be able to produce similar toughening because of their zero extensibility. The inhomogeneity of the system transforms previously formed large brittle cracks to a large number of small crazes, recognizable by the whitening of the stressed regions. A good adhesion between the elastomer particles and the matrix resin is essential. Crazing will constitute the mode of failure for brittle matrix resins including high-impact polystyrene. More ductile resins such as the acrylonitrile–butadiene–styrene copolymers will also deform by shear yielding.

A composite material should be recognizable by its optical opacity, whether both components represent optically clear polymers or not. This rule, however, cannot always be applied since not too long ago good optical clarity has been obtained when either the particle size of the dispersed phase is too small or all components' refractive indexes are the same. Invariably a whitening will be observed when these composites are stretched beyond their yield points since more pronounced phase separation is likely to occur.

There are several possibilities for obtaining these kinds of composites, but the actual process used is generally not divulged by the manufacturer. In the simplest cases two separately polymerized materials are blended. If the components are miscible, the resulting plastics can be related to plasticized materials. If they are immiscible or only partially miscible, they would be considered in the broad sense to be composites. Commercially available alloys or blends could belong to either group.

Thermoplastic matrix materials dominate several properties of composites, such as solubility in solvents, heat deflection temperature, and processing conditions. The tensile strength and modulus of elasticity are influenced to a lesser degree. The elastomeric dispersed phase will mainly affect the material's elongation and impact properties. But the two components can also be polymerized together, for instance by (co)polymerizing one phase first and then letting the other phase grow as a block-graft (co)polymer, branching off the original (co)polymer chain. The reader is referred to Chapter 5.

FIBER-REINFORCED PLASTICS

Fiber-reinforced plastics utilize the high tensile strength and high modulus of elasticity easily obtainable with oriented fibrous materials. The wide spectrum of properties realizable should be characterized with examples covering each extreme:

1. The intermediate tensile strength plastics, which contain randomly distributed fibrous filler materials, and
2. the precision wound high modulus fiber filament wound structures.

Any fiber material can reinforce any plastics matrix as long as the fiber has an inherently higher modulus of elasticity and ensures a good interfacial bond between fiber and matrix. In many instances surface treatment of the fiber is required to comply with this stipulation. But reports indicate that in unidirectional long fiber structures excellent results can be obtained when the matrix material keeps the fibers in position laterally but lets them slide in the length direction within regions where excessive strains may develop locally.

FIBER MATERIALS

For the selection of *fiber materials,* the following criteria must be taken into consideration. In most cases selection will start with the definition of the mechanical and thermal requirements placed on the formed part. Furthermore cost and availability of the required fiber form (short or continuous fibers) and the ease or difficulty of the fabrication process are of importance. When a high molding pressure and high flow length are involved, inorganic fibers may cause abrasive wear of molds.

In general, the available fibers are classified according to their tensile strength and their modulus of elasticity. Since these values are determined only in the direction of the fiber, highly oriented fibers, such as the high modulus carbon fibers, display a nonisotropic behavior that may become expressed by a relatively low impact resistance. This is based on their relatively lower shear strength. The completely isotropic glass fibers, in which none of the atomic bonds can be oriented in the fiber direction, manifest a supremacy in this regard.

Because of the small diameter of the fiber the differential in the coefficient of thermal expansion induces bond failure to a lesser extent than is experienced with large inserts in molded plastic parts. The coefficient of thermal expansion in the fiber directions may become negative because of the greater matrix expansion in the transverse directions, which causes greater undulation of the fibers in cross-plies. This occurrence is paralleled by the observation that a woven fabric or rope will contract when it absorbs water even though the individual fibers will extend lengthwise.

Of greatest importance for all fiber-reinforced composites is the orientation of the fibers in the formed part to withstand the expected imposed stresses during the life of the part. This makes it necessary to determine the magnitude of the stresses and to select a molding process that can provide the required concentration of fibers in those respective directions.

In low cost cellulose fiber-reinforced phenol- and urea-formaldehyde resins the relatively short fibers are randomly dispersed and during compression molding only a minor orientation of the fiber direction planar to the mold surface will occur, predominantly at the outer skin of the molded part.

The short glass fiber-reinforced thermoplastic resins allow only a relatively small concentration of glass fibers (10–30 wt%), otherwise processability by injection molding suffers. A slight alignment of the fibers will occur in the direction of flow during mold filling.

In unidirectional composite structures, the highest fiber concentrations are obtained by pulltrusion or by filament winding processes. To maximize fiber volume, the reinforcing fibers are applied as rovings or preimpregnated tapes. High fiber concentrations (up to 60 wt%) and good adaptation in regard to fiber orientation can also be obtained with unidirectional or bidirectional glass fiber-reinforced compression-molded laminates. Their use, however, is restricted to rather flat sheets. Unsaturated polyester and epoxy resins are still dominant in this field. Since the tensile strength of bidirectional laminates might be reduced to less than one-third when tested in the direction just between the two basic fiber orientations, quasiisotropic laminates are frequently preferred and obtained by laminating the reinforcing fibers in more than two directions (e.g., at 0°, 60°, and −60° or at 0°, 45°, 90°, and −45°). The anisotropy in tensile strength of glass fiber-reinforced laminates, as obtained by measuring tensile strength at varying angles from the principal fiber direction, is shown in Figure 13.1. Still, high strength values are obtainable only for two-dimensional structures. Identical compositions molded into heavy wall section parts, where an increasing fraction of the fiber reinforcement is distributed randomly and nondirectionally in all three dimensions, can obtain strength values lying only within the area marked in gray.

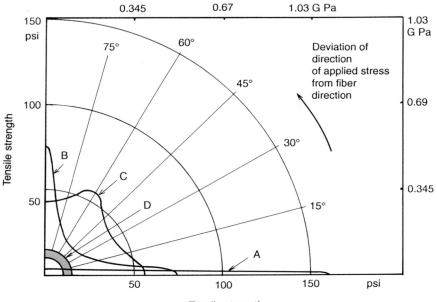

FIGURE 13.1. Anisotropy in tensile strength of glass fiber-reinforced composites. (A) Unidirectional fiber-reinforced materials (fishing rods). (B) Bidirectional, two-dimensional fiber-reinforced materials (flat glass cloth reinforced structures). (C) Three-directional, two-dimensional fiber-reinforced materials (quasiisotropic, staggered structures). (D) Random, three-dimensional fiber-reinforced materials (fiber filled injection or compression molded, heavy wall thickness parts). Adapted from Designing with Composite Materials. The Institute of Mechanical Engineering, London, 1973.

For irregularly shaped parts, sheet molding compounds have been introduced that provide good drapability as a result of the presence of randomly oriented chopped fibers or swirled continuous strand glass fibers and a filler system. These prevent the resin phase from rushing ahead of the glass fibers during distribution of the compound as it is pushed into distant parts of the mold. Glass fiber concentrations are only in the 30 wt% range, which is still higher than the less than 20 wt% glass fiber contained in bulk molding compounds.

Since better mechanical properties can be obtained with glass fiber-reinforced preform molding processes, the originally labor-intensive processes are now regaining interest. This is the result of their adaptability to automated procedures in which sophisticated robots are able to produce not too complicated three-dimensional parts for high-volume applications. A better distribution of fibers can be obtained either by blowing chopped glass fibers onto a suitably shaped metal screen with an air gun or by thermoforming a sparsely binder-impregnated continuous-strand fiber mat. In the first case air suction behind the screen holds the fibers in place until the small amount of binder, which is applied at a later time, is dried. Attaching to the screen a fiber-length high ridge at the perimeter of the part mitigates the need for trimming the preform and provides additional reinforcement at the perimeter. Glass fiber concentrations can attain a 40–60 wt% range.

Advanced Composites

Advanced composite materials are often differentiated from ordinary fiber-reinforced composites in a manner similar to the differentiation between engineering thermoplastics and commodity thermoplastics. In both cases no universally clear division has been accepted. In general, advanced composites contain higher modulus fiber reinforcements than the regular E-glass fibers and matrix resins with better mechanical but superior thermal properties than the conventional unsaturated styrene-modified polyester resins.

Since the covalent chemical bond potentially represents the strongest bond, surpassing metallic and ionic bonds, covalently bonded materials have been scrutinized with interest for their ability to form fibers. To obtain high specific stiffness and strength, lightweight elements have gained special attention. The carbon atom emerges as an ideal element; diamond indeed is the strongest, stiffest, and hardest material known. The drawbacks are that it is not available in fiber form and that it is very brittle. Other similar, covalently or nearly covalently bonded materials (e.g., boron, boron nitride, and silicon carbide) also represent high strength materials. Still, a conversion to thin fibers to overcome the inherent brittleness of these materials is required. This also minimizes the severity of failure-inducing flaws in these materials. Only in recent years has it become possible to obtain fibrous carbon or graphite in large quantities by the conversion of carbon atoms containing precursor fibers at high temperatures. The materials with the highest carbon content (99.9%) yield the stiffest fibers. These high modulus carbon fibers are, however, more subject to brittle failure modes during impact and shear loadings. It has been found that a hybrid construction, applying carbon fibers in conjunction with aramid or S-2 glass fibers, will significantly improve impact resistance.

The structure of these fibers is not clear, but one should envision those fibers as consisting of parallel tapes or rolls with the truncated benzene rings mainly lined up closely in the lengthwise direction. A possible similarity to an asbestos fiber structure has been pointed out in Chapter 2 (pp. 13–14). The extent of undisturbed covalent bond alignments of these benzene rings in one straight row makes the greatest contribution to the modulus of elasticity. (Straight does not refer to the mathematical straight line but to a corrugation of the carbon atom bonds with only every second atom being positioned on a straight line.) This pattern can be found in the diagonals of the diamond unit cell, in the polyethylene crystallite, and in the condensed benzene ring compounds. A somewhat less ideal arrangement of the carbon atoms can be found in the aromatic *p*-aramid synthetic fibers (Figure 13.2C), having about half the tensile modulus of the carbon fibers, a much lower temperature capability, but equal tensile strength.

Cumulated benzene ring structures do not necessarily have to be arranged in flat platelets, as is the case for regular graphite. Recently the structure of the buckminster fullerene, a molecule containing only 60 carbon atoms, has been identified. In this structure 20 benzene-type hexagons are wrapped around so they resemble a soccer ball. Each of the hexagon rings is surrounded by three other hexagons with the gussets in between forming 12 pentagons. These molecules are present in soot and no mechanical strength can be expected from such structures. However, in a ribbon-like arrangement of hexagons, a pleated or rolled shape of sheets should result in high strength materials comparing well with that of graphite or carbon fibers. As seen in Figure 2.9 (p. 15), such circular or spiral structures have unexpectedly been identified for asbestos fibers.

If the polymer molecules in polyethylene become properly aligned, as is now possible by the gel-spinning process, an extremely high modulus of elasticity in the fiber direction can also be obtained but their thermal capabilities remain limited to that of ordinary polyethylene, about 200°F (100°C). Figure 13.2 shows how chemically different compounds can become quite similar in regard to their macroscopic and microscopic structures. The structure of other strong fibrous materials, such as cellulose, collagen, and perhaps metallic whiskers, should also be visualized in a similar fashion. Of a different nature are the amorphous or randomly crystallized and also nonoriented drawn fibers such as those of glass, ceramic, and metallic materials.

Since most fields in which the application of advanced composite materials are being considered relate to mobile uses, the final weight of such contrivances is of great consequence. Therefore comparisons of different materials should not be made along volume-based properties—as is usually done—but by comparing load-carrying ability and deformations for equal weight structures. The specific strength properties shown in Figure 13.3 have been obtained by dividing the modulus of elasticity or the tensile strength, respectively, by the specific gravity of the tested materials.

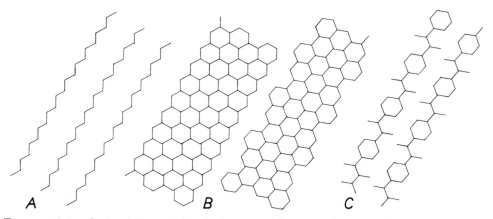

FIGURE 13.2.   Quite different chemical compositions can form similar fibrous structures of high modulus fibers from (A) gel-spun polyethylene, (B) carbon, and (C) *p*-aramid.

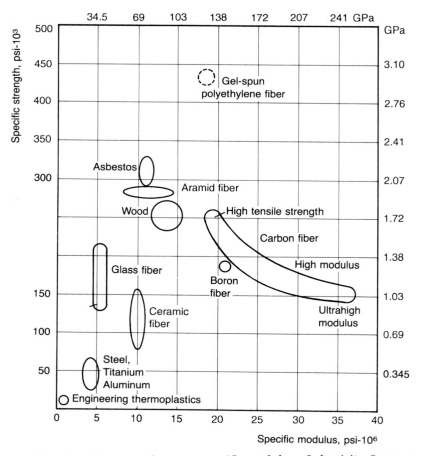

FIGURE 13.3. Specific strength versus specific modulus of elasticity for some strong materials. Adapted from M. Jaffe in Polymers: An Encyclopedia Source Book of Engineering Properties. John Wiley, New York, 1987.

There is no question that the high-strength *fibers* occupy the top positions in the right half and upper half of the graph. These values represent peak values, which must be multiplied by 0.6 to arrive at the strength values for their respective *unidirectional composites;* this is because they consist of only about 60% fiber and 40% matrix resin, which does not contribute to the modulus of elasticity of the composite. A further reduction in strength and modulus takes place when the desired structure is subject to stresses in other directions though in the same plane. Since practical laminates have only part of their fibers effectively aligned in any one direction, the obtainable strength properties will lie, depending on the makeup of the laminate, on a straight line to be drawn between the point marking the fiber's properties and the point of origin at the left lower corner. Quasiisotropic laminates are positioned about one-third up that line.

When all these values are compared to some other established materials, one must consider that in the case of metals, an equal strength extends in all three-dimensional space directions benefiting bolted, riveted, or welded fabrications. The three important aeronautical metals, aluminum, stainless steel, and titanium, all appear in close proximity in the graph, because lower specific gravity just about compensates for a lower modulus of elasticity. The main discriminating property among them lies in their temperature capability. Since wood is structured like a composite, a typical value for it has also been added to the chart. In regard to deflection or load-carrying capability, no material surpasses the specific strength properties of a solid, nonjointed piece of wood at ambient temperature.

Although the *matrix resins* do not contribute to the peak properties of the composites, they are of great significance when these properties must be maintained under adverse conditions—high temperatures, after exposure to moisture and chemicals—or when subjected to load endurance and fatigue. Cracking of the matrix material may be caused by deep thermal cycling (especially at low temperatures) because of the differences in the coefficient of thermal expansion, with carbon fibers in laminates having a near zero value. Lower modulus of elasticity matrix resins will be less susceptible to that failure mode. Also all the measures to improve and maintain good adhesion between matrix and fibers are important.

Thermoset resins still dominate the plastic matrix resins. They provide good bonds, high temperature creep resistance, and generally good chemical resistance. However, they are more difficult to prepare. For high-temperature advanced composites the preference for matrix materials is shifting to some very high-temperature-resistant thermoplastic resins, especially those of the semicrystalline or liquid crystal type. Because of faster production cycles this will lead to lower cost composite structures having very good high-temperature creep resistance. Furthermore they facilitate better fabrication, bonding, and repairing processes.

Given their distinct cost advantages, there should still remain sufficient demand for the two long established high production volume glass fiber composites: the reinforced unsaturated polyester resins, which represent the thermoset plastics variety, and the injection moldable short glass fiber-reinforced engineering plastics, which represent the thermoplastic variety.

CELLULAR PLASTICS

Cellular plastics or *foamed plastics* can be considered composite materials since they consist of two distinct phases. Their applications extend mainly over three areas:

1. structural foam moldings,
2. thermal insulation, and
3. cushioning materials.

Nearly all plastics materials, both thermoplastic and thermoset, are available as foams, encompassing densities ranging from 0.1 to 50 lb/ft³ (0.0015 to 0.8 g/cm³). These low densities are obtained by the incorporation of gases or vapors of low boiling liquids into the resinous materials. Gases such as nitrogen and carbon dioxide are injected into either the plasticized plastic or the liquid resin just prior to forming of the molded part. In other cases in which a chemical blowing agent is admixed into the molding compound, these gases will form when they chemically decompose at elevated temperatures (see Chapter 12, p. 277). Some other concurrently formed chemical fragments might, however, adversely affect the plastic's properties. Hydrogen gas liberated from sodium borohydride has a very high gas permeability, which will sharply reduce the degas time for parts that must be painted expeditiously while avoiding blister formation.

In the case of polyurethane foams a unique gas-forming method can be employed. Carbon dioxide will form when water is added to the catalyst (1). On the other hand, water must be strictly excluded when processing bulk urethane moldings or castings to prevent the formation of bubbles in the part.

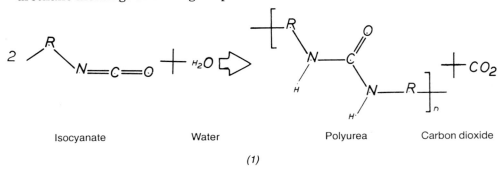

| Isocyanate | Water | Polyurea | Carbon dioxide |

(1)

The incorporation of low boiling liquids such as chlorofluorocarbons (halocarbons) or low-molecular-weight hydrocarbons or ethers into the plastics represent the preferred way to make foamed products. As liquids they are in many instances soluble with either the plastics or the reactive liquid components. In some cases such as the expanded polystyrene foams these compounds are shelf-stable for a prolonged time. Finally the shaped foam product is obtained, after heating the compounds by external means such as steam in the case of expanded polystyrene foam or when exotherm heat is generated during cure of thermoset formulations. In both cases the heat converts the liquid foaming agent to a vapor. These vapors will be exchanged by air in a relatively short time because of diffusion of the vapors through the cell walls of even closed cell foams. In other applications these vapors are retained in the foam to secure the best possible thermal insulation properties. This is especially important for refrigeration uses. Some chlorofluorocarbons occupy a most desirable position since they not only possess a very low thermal conductivity but also by far the lowest vapor transmission rates. Equally good replacement compounds are now being developed since the established chlorofluorocarbons are implicated in stratospheric environmental interferences.

A large number of foam molding processes are known. Selection is guided by the type and chemistry of the plastic under consideration, the application of the foam, and the desired shape of the foamed part. The key properties of the three categories referred to in the introduction should now be considered.

Structural Foam

Structural foam molding processes are preferred for processing very large and heavy cross-sectional parts since they provide several advantages over conventional injection molded items. A much stiffer part on an equal weight basis can be obtained and sink marks and warpage can be eliminated. Because of the greatly reduced injection pressure (about one-tenth) for complete filling of the cavity, much larger parts can be molded on smaller machines and with lower cost molds. Disadvantages that must be considered include the inferiority of some surface properties of structural foam parts. Surfaces are more easily indented or scratched than solid moldings and a smooth surface appearance without the typical swirl pattern can be obtained only with additional costly process variations. Furthermore, cycle time will most likely be extended as a result of longer cooling time requirements. The lower thermal conductivity of the foamed material and the retention of high-pressure gases in the center of heavy section parts may cause swelling when temperature is not sufficiently lowered while still in the mold.

Thermal Insulation

For *thermal insulation,* foams having the lowest density and a uniform small cell size perform best. Their structural integrity just has to withstand the sometimes rough installation conditions and endure shear forces that may arise, especially when used for mobile application. It is important to realize the limits imposed by the upper temperature capability of the various plastics materials since foams can deteriorate chemically at elevated temperatures much faster than the solid plastics of which they are made.

Cushioning Applications

Cellular plastics for *cushioning applications* occupy an intermediary range in regard to density. Their flexibility can range from strong and rigid crash pads to soft, resilient, and lightweight pillow fillings. The one property of rubber foams that cannot be easily duplicated by plastics is the high resiliency of natural rubber. Even the softest plastic foams manifest a relatively high stiffness during the initial deformation stage and also lack rubber's vigorous rebound force during unloading. They also are more temperature sensitive and exhibit much higher hysteresis losses in comparison to rubber foams. This manifests itself as an advantage when vibration isolation or sound deadening becomes desirable.

Biogenic Composites

Biogenic composites have not been given adequate recognition by chemists or engineers involved with the manufacture and application of composites. There might be some valid reasons:

1. They work best only under moisture saturated conditions.
2. They are designed to perform only within a narrow temperature range.
3. The formation of the polymeric materials—although massive in quantity—is too slow for many industrialized manufacturing processes.

On the other hand the locating and spacing of small atoms or the placement of even larger macromolecular structures—as so common for biogenic products—could represent a model for sophisticated constructions of some high-tech composites. From the vast number of biogenic structural materials, only two have been selected for in-depth analysis.

Cellulose

Cellulose represents the dominant structural compound found in plants. The monomeric unit of this material, which is generally described as a thermoset polymer, consists of two distinctly connected, water-soluble glucose sugar molecules. It is not known how the polymerization process develops but the end product is a water-insoluble, nonmelting, stretched out polymer chain consisting of a few thousand six-membered rings connected by oxygen atoms as an acetal bond. A great number of compounds could be constructed in very similar fashion. A number of them occur in biogenic tissues, but cellulose singularly exhibits the superior structural properties. In all technical books the basic chemical formula for cellulose is stated but the essential structural intricacies are not illustrated convincingly. The two six-membered rings can occur only in two steric conformations as known from the two cyclohexane structures. The zigzag or "chair" form is somewhat more stable than the "boat" form. Both rings in cellulose and other similar natural structural products exhibit the "chair" form. That one of the ring members in cellulose is an oxygen and not a carbon atom does not alter this consideration. The important features in conformation are noticeable only when observed in a perspective view.

The arrangement of the ring atoms is akin to that found in diamond (see p. 47). By locating the cyclohexane-like rings in the diamond crystal, it is apparent that in all cases the other two remaining bonds at each ring carbon atom either extend exactly vertically as axial bonds or spread out sideways inclined, continuing the undulation as zigzag lines, as equatorial bonds. Figure 13.4 illustrates these two types of bonds for cyclohexane and for the two glucose molecules, α-glucose and β-glucose. The bonds that will participate in the polymer chain are highlighted.

It is important to realize that in the case of β-glucose all axially extending bonds are occupied solely by hydrogen atoms. The hydroxyl groups and the hydroxymethyl group spread out in near equatorial directions only. The angu-

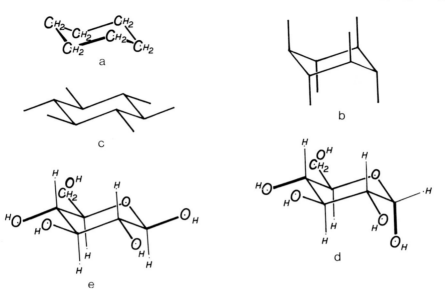

FIGURE 13.4.   Directions of chemical bonds in cyclohexane and glucose. Bonds which will form the polymers are highlighted. (a) Cyclohexane; (b) six axial bonds; (c) six equatorial bonds; (d) α-glucose; (e) β-glucose.

lar positioning of the two hydroxyl groups at the right and left extremes of the α-glucose molecule already clearly indicates that this monomer is unable to form a straight crystallizable polymer. The two glucose "dimers" of importance, cellobiose and maltose, which represent the first polymerization steps leading to cellulose and starch (amylose), respectively, are shown in Figure 13.5. The two rings in the cellobiose molecule can almost be drawn in one plane—not possible for the maltose molecule.

Only in the case of cellulose can the polymer chain follow a continuous zigzag line in an overall straight direction. Because of the free rotation at the acetal oxygen bonds, consecutive glucose rings are flipped 180° and then stabilized by the strong hydrogen bonds connecting them to neighboring chains in a crystallite arrangement basically similar to that exhibited by other polymer crystallites. In the case of amylose (starch) that regular zigzag line is interrupted at every first carbon atom of the following glucose ring because of the two sequential zig and zag bends. Here sequential rings cannot be positioned in one plane, no matter how they are rotated around the acetal bond. Therefore it is assumed that the starch molecule may form a spirally wound, more random, conformation. The amylose molecule can form crystalline regions only when sufficient water molecules occupy the free space and stabilize the curled chains with bridging hydrogen bonds. Nonstructural properties of these types of compounds are exploited in dextrin and cyclodextrins.

To further emphasize the essential features, only the ring carbon and oxygen atoms with their oxygen link atoms of these two polymer chains were first projected onto a flat strip of paper. After the appropriate bends were folded in

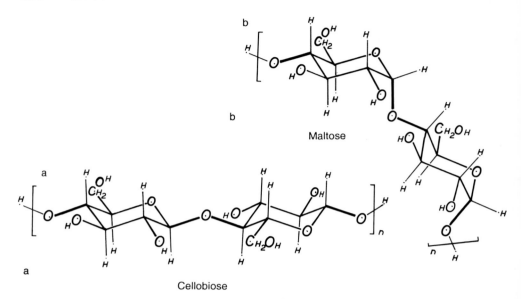

FIGURE 13.5.   Chemical structure of cellobiose (a) and maltose (b) with brackets marking the locations at which molecules polymerize to cellulose and starch.

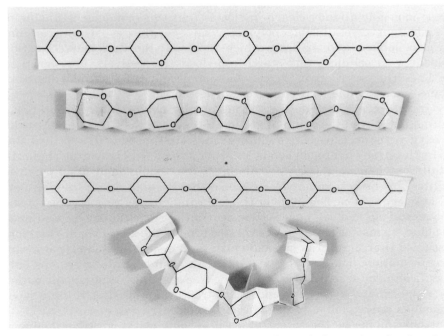

FIGURE 13.6.   Conformation of cellulose (top) and starch (bottom) molecules. Only the atoms that form the polymer chains are shown.

the paper strips and some relief provided at the oxygen links, the resulting structures were photographed as shown in Figure 13.6. The difficulty of expressing the structure of randomly arranged polymers is also encountered with other materials such as the phenol-formaldehyde resins and rubbers.

The unique location of the three hydroxyl groups in the cellulose molecule confers to cellulose materials not only their strength but also their chemical stability. This includes being insoluble in nearly all solvents and solutions and having the best thermal stability of all carbohydrates (compounds consisting only of carbon, oxygen, and hydrogen with atomic ratios of equal numbers or a few less oxygen atoms than carbon atoms but always two hydrogen atoms for each oxygen atom).

An important fact that can enhance our understanding of many semi-crystalline synthetic polymers is illustrated by the following example. A growing plant should have the ability to attach each glucose or cellobiose molecule to a growing cellulose micelle in any (crystal) order until the molecule reaches either the end of the intended length (up to 10,000 glucose units in cellulose) or until it borders a cell wall. An exclusively stretched-out arrangement would provide a material with an extremely high modulus of elasticity but would also make it much too vulnerable to fracture under high bending forces. Instead, one molecule, although maintaining its general orientation, will participate in hundreds of small crystalline regions for a stretch of only about 10 nm, which is comparable to synthetic polymer crystallites. The degree of crystallinity of cellulose in wood (about 60–70%) is also comparable to many semicrystalline plastics. Chain folding of cellulose molecules in micelles has not been seriously proposed by cellulose chemists. Plant tissues should have no problem—if that would be the preferable or desirable mode—in achieving regular chain folding by utilizing their ultrastructure processing ability found in all biological systems.

As a curiosity it should be mentioned that in earlier times oddly shaped parts of trees such as crooks, knees, and forks have solved many joining or shaping problems for craftsmen. The difficulty of efficiently joining wooden members, one of the main reasons why wood is limited in many potential applications as a structural material, is effortlessly overcome in biogenic structural systems. The arrangement and orientation of micelles in microfibrils are quite adaptable, so that apparently abrupt changes in directions as they occur from the stems to the branches and to the roots can reliably be handled by plants.

A simple but time-consuming manufacturing process has been utilized to produce superior wooden pitchforks (Fig. 13.7) by positioning selected thin twigs of branches into desired positions and letting the plant supply the structural polymer at the most advantageous positions and orientations.

Although the same cellulose molecule is universally present throughout the plant world, the microfibrils are very specific to each plant and plant tissue, and are therefore responsible for the multiplicity of the plant world. Of even more significance is the fact that growing trees are capable of adding cellulose fibrils to existing tree branches in such a state that the polymer molecules are already under strain. This is known as tension wood in hardwood and compres-

FIGURE 13.7.   Wooden pitchfork produced by a natural manufacturing process.

sion wood in soft wood. If that were not the case, side branches would bend downward because of the increased weight added yearly (see Figure 6.27, p. 180).

Figure 13.8 shows the arrangement of the cellulose molecules in the crystalline cellulose micelles. The corrugated ribbons are all parallel side by side with alternating layers transposed halfway between and shifted in the lengthwise direction so that the pending hydroxyl groups can hydrogen bond with oxygen links, interconnecting layer to layer. This tight packing is reflected in the differences of the specific gravities of cellulose (1.53 g/cm³ for the amorphous regions and 1.625 g/cm³ for the crystallites) versus that of starch (1.46 g/cm³).

Although the cellulose molecules cannot be made visible in the electron microscope, the texture developing from the aggregation of micelles containing several hundred molecules can be discerned. The orientation of the cellulose molecule remains in the length direction of the micelles and those in turn are similarly oriented in the microfibrils, but more liberally arranged, forming the macrofibrils and the secondary wall of a plant fiber extending

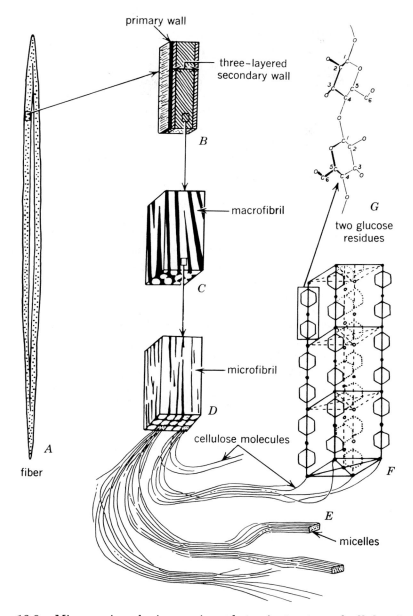

primary wall

three-layered
secondary wall

B

macrofibril

C

G

two glucose
residues

microfibril

D

cellulose molecules

A

fiber

F

E

micelles

FIGURE 13.8. Microscopic, submicroscopic, and atomic structure of cellulose in plants. Fiber (*A*) has a three-layered secondary wall (*B*). In a fragment of the central layer of this wall (*C*) the macrofibrils (white) consist of numerous microfibrils (white in *D*) of cellulose interspersed by microporosities (black) containing noncellulosic wall materials. Microfibrils consist of bundles of cellulose molecules, partly arranged into orderly three-dimensional lattices, the micelles (*E*). Micelles are crystalline because of regular spacing of glucose residues (*F*). These residues are connected by β-1, 4-glucosidic bonds (*G*); hydrogen atoms not shown. Courtesy K. Esau, Plant Anatomy, 2nd ed. John Wiley, New York, 1967.

over several inches. This illustration summarizes the relationships in regard to length dimensions from the cellulose molecule repeating unit which is 1.03 nm in length to a fiber a few inches long. Each step reflects a hundredfold increase in size. In wood and cotton the microfibrils are, in addition to being straightened out in the growth direction, also spirally wound at various angles in successive tubular layers.

The long cellulose molecules bestow tensile modulus and strength to the fiber. To also provide flexibility to the fiber, the stiff crystalline micelles must be interrupted and surrounded by some noncrystalline regions in which a matrix substance holds the cellulose micelles in position in areas where flexural and shear forces occur. Both lignin and the hemicellulose polymeric materials are mentioned in the literature as matrix materials for wood. These constituents must be removed before the cellulose can be subjected to any chemical processing.

The structures of these true composites are further expanded not only to improve the specific strength of wood but also to provide vessels that are so important for the transport of water and nutrients for the plant. This loosening of the structure results in the relatively low specific gravity for the various types of wood, primarily between 0.35 and 0.7 g/cm^3.

Since several cellulosics—plastics derived from natural cellulose by chemical conversions—have entered the broad field of thermoplastics, their relationship to the thermoset cellulose must be explained. As pointed out before, native cellulose has been polymerized and shaped simultaneously into a thermoset material in various shapes as manifested by the multifarious kinds of plant structures. To convert this material into other, more desirable shapes it must be cut into smaller pieces and rejoined with mechanical fasteners or adhesives.

Another method consists of chemically dissolving the crystalline cellulose micelles with the help of aggressive chemicals that can overcome the strong hydrogen bonds originating from the three hydroxyl groups of the glucose units and reconstituting the cellulose substance by neutralizing those chemicals.

To understand this process, one must realize that the crystalline regions of cellulose are interspersed with amorphous regions. The latter represent those regions in wood that are swelled up by water in their natural state, providing the great flexibility to "green" twigs and lumber. Well-dried wood will partly lose that flexibility. The amorphous regions are also responsible for the unwanted dimensional changes experienced with wood when it is subjected to varying relative humidities for an extended period. Some of the acetal bonds of the cellulose polymer in the amorphous regions might hydrolyze preferentially when subjected to chemical reactions, reducing the high molecular weight of cellulose significantly.

The regenerated cellulose, though chemically identical to the native form, consists of a cellulose of a much lower degree of polymerization and no longer possesses the high ratio of fairly well-ordered crystallites but a more amorphous, disaggregate structure with less orientation, more affinity to water, and higher dyestuff absorption. This also causes the lower modulus of elasticity and strength of these products, mostly fibers (rayon) and films (cellophane), as well as their lower chemical and thermal resistance.

On the other hand, purified and dissolved cellulose can be chemically reacted so that most of the three hydroxyl groups become capped. Because of the loss of the hydroxyl groups, the cellulose molecule can no longer partake in hydrogen bonding and therefore loses its thermoset character. These solubilizing processes must be well controlled since each of the various cellulosics requires a particular degree of polymerization to obtain optimum melt or solution viscosities for processing. Some of them may be soluble in water and others in organic solvents. The first ones represent the water-soluble cellulose methyl ether, carboxymethyl cellulose, and hydroxyethyl cellulose. The organic solvent-soluble cellulose esters are obtained by combining prereacted cellulose with acetic, propionic, and butyric acid. The latter polymers are thermally processable like other synthetic thermoplastics. Although these thermoplastics have lost their crystallinity, they possess rather good impact properties because of the random orientation of their cellulose molecules. They must be regarded, like ordinary thermoplastics, as isotropic bulk molding materials— though mostly compounded with plasticizers—and not related to composite materials anymore. Peak properties can be obtained by orientation when subject to the various fiber-forming processes.

Collagen

The second example of biogenic composites, *collagen,* represents one of the highest tensile strength materials in the animal world. It is the strength-providing material in tendons, sinews, ligaments, muscles, and hides and a minor component in bones enabling them to resist tensile and shear forces.

Cellulose represents the chief structural material in plants and provides all the static strength needed; however, the structural requirements in the animal world are quite different. To obtain mobility only articulated levers can be used. The concept of wheels is an impossibility in nature. The bones, consisting primarily of crystalline inorganic mineral compounds, represent the rigid components in that system and the high tensile strength tendons and ligaments represent the highly flexible connections between muscles and bones or bones and bones, respectively. Therefore not surprisingly, the molecular structure of collagen is quite different.

Like a rope formed from three tows, the triple-stranded structure of collagen is obtained by twisting three polymer chains, each consisting of the recurring three simple amino acids glycine *(2a)*, proline *(2b)*, and hydroxyproline *(2c)* forming the polyamide chain shown in *(3)*. The resemblance of this "copolymer" to the thermoplastic polymides is illustrated by two truncated formulas *(4a,b)* in which only the chain link atoms are contained. Since each of the side groups (only hydrogen in the case of glycine) of these amino acids rotate just under 120° to the left around the fiber axis and since interchain hydrogen bonds

connect the $\diagdown$C=O and $\diagdown$N—R links, the basic triple helix collagen fiber

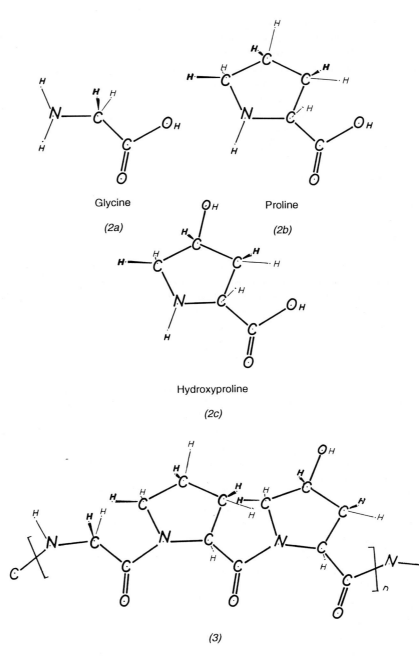

Glycine

(2a)

Proline

(2b)

Hydroxyproline

(2c)

(3)

exhibits a slight right-handed twist. The slender glycine monomer is always most centrally located in the triple helix (Figure 13.9).

As in cellulose, the coherence of the molecular chains is ensured by the correct positioning of secondary bond donors and receptors along the chain at

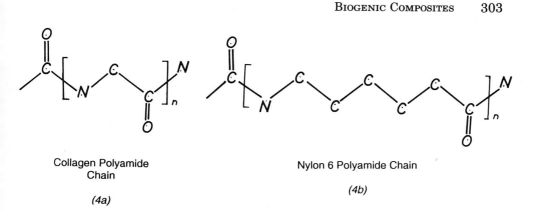

Collagen Polyamide
Chain

(4a)

Nylon 6 Polyamide Chain

(4b)

exactly matched staggered intervals. This structure is a very good illustration of the fact that an orderly structuring is necessary to obtain a high tensile strength material, but crystallinity is not. Of importance is that the polymer chain links become positioned in an orderly fashion oriented in one direction.

The matrix material is of subordinate importance in this case since, unlike other composites, stiffness or rigidity is not required. Nonspecific polysaccharides (chondroitin sulfate) may serve to hold the fibrous collagen strands in desirable positions. For materializing hard and stiff substances such as bone and teeth, inorganic crystalline materials combine with collagen in a manner similar to interpenetrating composites. More knowledge about these important biological mineralization processes may also promote developments for industrial composites.

The arrangement of successive monomer units in helix form is also quite abundant in synthetic polymers. Unlike biogenic polymers in which helix formation is a prerequisite for obtaining a desired performance, in synthetic polymers the helix formation is the result of steric hindrances originating from chain ligands. In the case of polytetrafluoroethylene just a slight twist is required (13 monomer units per full turn) to accommodate the larger fluorine atoms when contrasted with polyethylene. Bulkier substituents will result in much shallower helixes. The isotactic polymers of polypropylene, polybutylene, and polystyrene require three monomer units per turn and poly-4-methyl pentene requires three and a half.

As in cellulose, the close steric arrangement of the monomer building blocks in collagen is limited to the native materials. If collagen is heated in water the secondary bonds are loosened and a viscous solution of the dispersed polyamides is obtained. On cooling some of the hydrogen bonds reform but since an ordering mechanism no longer exists, the formed triple helixes become fragmentary and disorderly. The solution converts to an intermolecular network or gel. Similar to regenerated cellulose, the collagen solution known as gelatin can still serve as a respectably strong material. Not too many years ago gelatin represented the preferred binder for paints, for formed parts and for both paper and wood glues.

Other natural polymeric materials such as some *proteins* are completely different biogenic materials. In these compounds the building blocks, consist-

FIGURE 13.9. Symbolized structure of collagen. Three polypeptide chains forming a helix. The two amino acids with rings parallel to their chains are marked by flags. The hydrogen atoms of the amino acid without side groups are connected with the other two polypeptide chains in the center of the rope by hydrogen bonds.

ing of more than a dozen different amino acids, are always added in a well-defined sequence and therefore their molecular weight is always constant. Also their spatial conformation is very specific for each kind [e.g., the folding pattern of the hemoglobin molecule and the double helix structure of deoxyribonucleic acid (DNA)].

Still unexplored is how to extricate some of the features of biogenic structural materials to promote further progress in the field of advanced composites. To duplicate the formation of the strong materials in plants and animals may still be too ambitious, since we lack knowledge of how living things sculpt their shapes. The chemist is still incapable of influencing the conglomeration of polymers to form suprastructures.

The origination of suprastructures in nonliving substances, including polymer crystallites, must be considered to stem from wave-type, oscillatory, or cumulative disturbances. Examples are the orange peel effect on paint or metal surfaces, the mud cracking in ceramics, the moiré fringes manifesting a certain, otherwise unnoticeable ordering, and the wave formations revealed in cirrus clouds, a body of water, sand, or a piece of cloth in the wind. In Figure 13.10 a sandy desert shows remarkable similarity to electron microscopic pic-

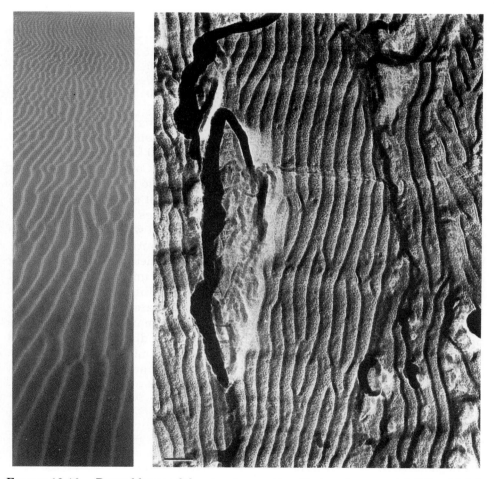

FIGURE 13.10.   Resemblance of the structure noticeable in polymer crystallites (right) with wave formations on sand (left). Electron micrograph of an extended chain fracture surface of polyethylene. From B. Wunderlich and L. Melillo, Morphology and growth of extended chain crystals of polyethylene. Makromol. Chem. **118**, 250, 1968.

tures of lamellae structures of polymeric materials. Applying the fractal concept explained in Chapter 6 (p. 149), their structure becomes increasingly comprehensible.

A much simpler synthesis of other natural structural materials could be regarded as being successfully duplicated by many of the spun synthetic fibers. Still, the intricacies of how an unstructured liquid solution is excreted and externally (separated from any life-controlled processes) formed into a strong structural material by means of a simple drying process (without any toxic by-products) are not all known. Astounding examples are the high strength fibers, the silk of the silkworm and dragline silk of the spider, and the wing formation of insects. The tensile strength and elongation for selected fibers are given:

| | Tensile Strength | | Elongation |
Fiber	ksi	MPa	(%)
Dragline spider silk	200–250	1400–1550	16–30
Polyaramid fiber	290	2000	4
High tensile nylon fiber	230	1600	16
Carbon fiber	250	1750	1

An example of great success is the tedious gel-spinning process of ultrahigh-molecular-weight polymers resulting in fibers with a tensile modulus of up to 15,000 ksi (100 GPa) and a tensile strength of up to 200 ksi (1400 MPa). These extraordinary properties affirm some important facts. Covalent bonds such as the —C—C— bond can provide very respectable tensile properties as long as the polymer chains are sufficiently long and a parallel alignment can be ensured. In Figure 13.2A this arrangement, which applies to the gel-spun polyethylene fibers in particular, is shown. When observed within the lonsdaleite (a diamond modification) crystal lattice, only every fourth row of all the carbon atoms arranged in a zigzag conformation becomes occupied by strong covalent bonds of the polyethylene chain. The crystal unit cell dimensions of polyethylene are remarkably similar to the corresponding dimensions of lonsdaleite, even though the nature of the bonds in both materials are quite different in their perpendicular directions. Therefore the maximum value for the modulus of elasticity of gel-spun polyethylene in the c direction (fiber) should be about one-fourth that of the diamond value of 165 million psi (1130 GPa). The top literature values of 34 million psi (240 GPa) are indeed quite close.

| | a direction | b direction | c direction (only C—C bonds) |
Crystal unit cell	(nm)	(nm)	(nm)
Polyethylene	0.74	0.49	0.253
Lonsdaleite	0.866	0.412	0.253

The relatively low tensile properties of the common industrial fibers are the result of

1. their insufficiently high molecular weight and
2. their imperfect alignment of the chains even in highly oriented fibers.

In the case of randomly arranged chains, as they occur in injection molded plastics parts, only a very small fraction of the chains are aligned in any one direction.

For the construction of advanced composites it is important to realize that parallel, straight fibers will always yield the highest modulus of elasticity. An increasing reduction in the listed order will occur when the same fibers are applied as twisted yarn, woven goods, braided or knitted fabrics, and nonwoven fabrics. From observations of biogenic composites (wood, insect cuticles, muscles and tendons, bone and teeth) one must conclude that the following conditions are of importance:

1. Exactly positioning and orienting the bulk of the strongest uniaxial material parallel to the direction of highest stress to achieve the highest stiffness and strength.
2. Additional positioning and orientation of strong uniaxial materials in a skewed framework (neither parallel nor intersecting) and interlayering of more isotropic or transverse reinforcements, stacked in narrow intervals to promote extensibility and toughness.

On the one hand, addition of some chopped glass fibers to a molding compound appears in this context to be quite dilettantish. On the other hand, presently applied computer-aided manufacturing processes, in which robotic equipment can supplant tedious handwork of precisely laid-up composites utilizing multiple reinforcing fibers (of varying modulus and toughness), can be seen as the closest approximation to biogenic structural materials.

REFERENCES

M.F. Ashby and D.R.H. Jones, Engineering Materials. Pergamon Press, Oxford, 1980.

J. Bodig and B.A. Jayne, Mechanics of Wood and Wood Composites. Van Nostrand Reinhold, New York, 1982.

B.L. Browning, The Chemistry of Wood. John Wiley, New York, 1963.

Designing with Composite Materials. Institute of Mechanical Engineering, London, 1973.

K. Esau, Plant Anatomy, 2nd ed. John Wiley, New York, 1967.

M. Grayson, ed., Encyclopedia of Composite Materials and Compounds. John Wiley, New York, 1983.

M. Feughelman, Mechanical properties of fibres and fibre thermodynamics. In Applied Fibre Science, F. Happey, ed., Vol. 1, p. 43. Academic Press, New York, 1979.

J.E. Gordon, The New Science of Strong Materials, 2nd ed. Princeton University Press, Princeton, NJ, 1976.

D.J. Johnson, High-temperature stable and high-performance fibres. In Applied Fibre Science, F. Happey, ed., Vol. 3, p. 127. Academic Press, New York, 1979.

G. Lubin, ed., Handbook of Composites. Van Nostrand Reinhold, New York, 1982.

C.H. Pearson, Collagen and procollagenes. In Applied Fibre Science, F. Happey, ed., Vol. 3, p. 411. Academic Press, New York, 1979.

R.W. Schery, Plants for Man, 2nd ed. Prentice-Hall, Englewood Cliffs, NJ, 1972.

M.M. Schwartz, Composite Materials Handbook. McGraw-Hill, New York, 1984.

R.B. Seymour and G.S. Kirshenbaum, eds., High Performance Polymers, Their Origin and Development. Elsevier Science Publishing, Amsterdam, 1986.

S.G. Shenouda, The structure of cotton cellulose. In Applied Fibre Science, F. Happey, ed., Vol. 3, p. 275. Academic Press, New York, 1979.

E. Sinnott, An Introduction to Plant Anatomy, 2nd ed. McGraw-Hill, New York, 1947.

14

Processing Methods for Polymeric Materials

Although in biological systems the formation of discretely formed shapes occurs simultaneously with the chemical buildup of the macromolecules, in all man-made structural entities both events are always distinctly separate. This is also true for the formation of plastic parts. The sequencing of events can take place in two ways. Either the polymer is first synthesized and then molded into the desired shape or a precursor substance is first cast or formed and then polymerized to obtain structural properties. Only occasionally, such as when spinning fibers out of reactive solutions, do these phenomena overlap. The interwoven process of growth, which is so characteristic for biological systems, is not likely to become imitable.

The number of processing methods for plastics has increased at an accelerating rate in past years. In many instances these processes are just variations of the basic methods of pressure molding, extruding, casting, and mechanical forming. This book has not included detailed descriptions of molding processes; only shape or property manifestations caused by certain processing steps have been covered. In this chapter processing methods for plastics will be treated in a clearly definable but noncustomary way.

CAVITY FILLING PROCESSES

The production of parts utilizing cavities represents the greatest number of applications, extending from compression molding, stamping, transfer molding, and injection molding to the various reaction injection molding and liquid resin casting processes. A mold cavity is generally bound by two mold halves, the surfaces of which will completely delineate the expanse of the molded part. In injection molded or cast parts only a very small area is left open for the entrance of the plastic. The positioning of these gates is important to secure filling of all mold segments, to avoid displaying the appearance blemishes (splay marks) in their vicinity, and naturally to attain the most advantageous properties of the part. Because all surfaces represent replications of the mold cavities, great freedom is given to the designer in regard to the shape of the part. There are also various ways to obtain more complicated parts, including the use of internal or external threads and undercuts, and the incorporation of metal inserts. The size of the part is limited only by the clamp pressure of the injection molding machine, which presently extends to about 3000 ton. This allows a projected area of about 10 ft² (1 m²) if the injection pressure is assumed to be 10,000 psi (70 MPa).

Since most molds must be made of hard and durable materials, their cavities represent a fixed volume to be filled with the melted plastic or liquid resin mixture. High injection pressures can be applied to ensure proper filling of cavities. Their preeminent advantage is that they make it possible to produce parts repetitively, necessitating a minimum of finishing operations and producing items of identical weights and identical dimensions in all directions.

Invariably a shrinkage must occur whether the plastic is solidified only by cooling or whether the resin is chemically reacted (polymerized) in the mold. The amount of shrinkage is affected by the chemical structure of the polymer and in particular by the extent the polymer will consolidate into a crystalline structure occupying less volume. Therefore molds cannot be universally reassigned from one kind of plastic to another if precise part dimensions are of importance. Processing conditions and mold parameters may also have an influence. Simple molds for the preparation of standard test specimens are usually too rudimentary, resulting in higher shrinkages. Production molds having many geometric features create obstacles to the free shrinkage of the polymer, but may lead to a higher probability of emergence of harmful residual stresses. In Figure 14.1 the two extremes are illustrated. A flat round disc can shrink unimpeded during solidification, whereas a similar disc with a rim remains restrained as long as it stays in the mold. Although volume shrinkage would be the same in both cases, A will shrink more in diameter, whereas B will shrink more in thickness.

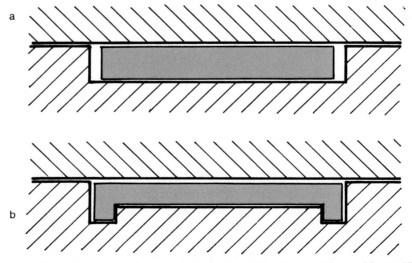

FIGURE 14.1.   Differences in shrinkage depending on the shape of the mold. (A) High shrinkage of molded flat disc. (B) Low shrinkage when provided with a rim.

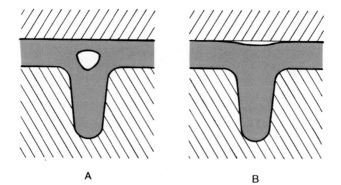

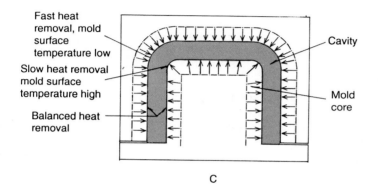

FIGURE 14.2.   Causes for (A) internal voids, (B) sink marks, and (C) warpage. (A) Void formation: plastic solidifying fast at mold surfaces. (B) Sink mark: plastic shrinking away from mold surface because of early solidification in the rib. (C) Warpage: arrows indicate cooling intensity.

The occurrence of shrinkage will lead to more serious consequences when it becomes inconsistent or irregular depending on the direction of flow. In heavy wall section parts, it might become manifested by sink marks or voids and in spread out, flat parts by distortions, warpage, and dimensional inaccuracies.

These disparities may be the result of several often interacting causes:

1. Abrupt and great changes in wall thickness. Although the thinner sectioned areas have already solidified, the heavy section areas still remain fluid, making it possible to replenish the volume shrinkage that took place in faster solidifying regions. This creates a large void where solidification occurs last (Fig. 14.2A).
2. Separation of the solidifying part surface from the mold surface when shrinkage starts will impede further heat diffusion out of the plastic. Those areas (Fig. 14.2B) become noticeable by the formation of sink marks.

3. Mold surface temperature disparities that become established between the surfaces of core and cavity areas of any mold. It is easier to conduct heat away from massive cavity parts of the mold than from the more compact core parts that have a much greater surface-to-volume ratio (Fig. 14.2C).

The reasons for warpage can generally be related to one of the four following causes:

1. temperature disparities in the plastic material caused by differences in mold surface temperatures or excessive flow length;
2. differences in packing pressure, again exacerbated by excessive flow length;
3. local variations in crystallization rates; or
4. flow induced orientation of fibrous fillers or polymer chains. The latter becomes especially pronounced with liquid crystal polymers or high shear processing.

Since an appreciable surface temperature difference may occur at the outside corners of mold cores, high heat conductivity beryllium–copper cores can improve productivity and alleviate some of the other problems mentioned. The 10-fold increase in thermal conductivity over steel molds still cannot completely compensate for the low heat diffusivity of all plastics, which still—in most cases—will dominate molding cycle times.

Other possibilities exist to avoid or reduce the severity of these problems. Adding distinct patterns to the mold surfaces and reducing the spread between the temperature of the mold and the temperature of the injected plastic are always beneficial.

Because of the easy expansibility of gases, structural foam molded parts will generally be free of these defects. The shrinking plastic is not pulled away from the mold surface since the interior gas bubbles will expand and still apply pressure even though the gate areas have already solidified. Similarly, sink marks along heavy ribs can be avoided by the various gas-injection molding processes. Before completion of cavity filling, a gas (usually nitrogen) is thrust into the mold so that the still liquid plastic in the center of the heavy ribs is pushed further to the mold extremities. The interior of all the heavy cross-section stiles forms a continuous gas pressurized channel. Because of an overall lower injection pressure and the elimination of all massive sections, these parts will also have lower molded-in stresses.

Similar results can also be obtained by a sandwich molding process. In this case, the injection of a plastic foam material takes place after the mold is partially filled with a strong skin material.

Newer more sophisticated injection processes have been developed for processing higher cost materials such as liquid crystal plastics where superb part properties justify higher production costs. By controlling and sequencing plastic flow directions, more isotropic properties are obtained. These processes, requiring multigate molds, are termed multiple live feed and push-pull injection molding.

Solidification of the plastic in cooled molds always starts at the surface. Since the material below the surface consolidates at a later time, the shrinkage that occurs will place the central regions under tension. To compensate for that force, the outer layer of molded parts must bear a compressive force. This represents one of the reasons why molded parts exhibit better mechanical and chemical properties than equally shaped machined parts. In turn, annealing of machined parts will have a similar beneficial effect on properties, the result of both the reduction of surface stresses and the cooling step, which will also put the surface under a compressive force.

The molding shrinkage of injection molded parts will generally be higher in the direction in which the material is flowing. This is because of the prevailing orientation of the polymer chains in the flow direction during the mold fill cycle. The shorter the molding cycle the less time the material has to randomize this orientation before solidification. This apparent discrepancy, however, should not be taken as a contradiction of the fact that the coefficient of thermal expansion is lower along the polymer chain direction than in the transverse direction. In glass fiber-filled plastics these fibers align in the same direction. Since the coefficient of thermal expansion of glass is so low, in this case the plastic shrinks considerably less in the flow direction. Warpage is still a possibility because of higher shrinkage in the transverse directions. Still, semicrystalline thermoplastics exhibit the highest overall mold shrinkages. The origin of dimensional digressions can lie in flow-induced stresses, in thermal stresses, or in orientation, often complicating the determination of the best remedy.

On parts with flat surfaces, the layer removal technique can help to identify the distribution of stresses in those profiles. By carefully machining off the surface to different depths, the extent to which selected segments of the part will bend indicates the stress imbalance in those cross sections.

A uniform distribution of fiber reinforcement is sometimes not attainable in composite moldings because the fibers diverge from the resin in parts where the plastic must flow through narrow channels or over ridges. Fiber breakage can occur in areas of higher fiber accumulation during compression molding. For all the above processes, the volume fraction of fibers must therefore be kept considerably below an otherwise desirable level.

Two effects prevail with fiber-reinforced molding compounds:

1. The weld lines, which form when, during filling of a mold, an advancing front reunites behind a pin or core become very weak because of the restriction of intermixing.
2. Since the reinforcing fibers are neither perfectly aligned nor ideally randomly distributed but oddly bunched up, the surface of the part shows both a fine surface roughness and a longer interval waviness, depending on the angle at which the fibers meet the surface. The incorporation of low profile or low shrink additives to the sheet molding or bulk molding compounds represents one remedy, though it is not completely effective.

To obtain mechanically stronger parts, a resin transfer molding process has been developed that utilizes reinforcements consisting of longer fibers and continuous swirl fibers in addition to the chopped fibers. This preformed fibrous structure is inserted into the mold prior to filling it with resin.

Compression molded and cast parts will produce much less warpage problems because of the lower shear forces applied during filling of the molds. Therefore an injection compression molding process is sometimes utilized for large shallow parts. This involves the injection of the required amount of plastic into a not completely closed mold. The mold is then closed to fill all segments of the cavity.

Still, considerably less flow will occur during the stamping or flow-forming process. A consolidated fiber-reinforced sheet needs only to be heated somewhat above the glass transition temperature before it can be formed in a cooled mold at high production rates, similar to sheet metal forming processes.

Continuous Forming Processes

In these forming processes the plastic must not exceed a relatively high viscosity state, at which it becomes just formable but remains stiff enough not to flow out of shape when—usually without cooling—it emerges out of the forming die. Higher molecular weight polymers must be used for extrusion and calendering processes to prevent sagging or distortion of the formed product. In all these cases the plastic is never completely encased in a mold but only transiently restrained. The plastic is continuously fed into one side of a zone where forming takes place and exits at the other side roughly in the desired shape. A variety of processes embody extrusion forming: sheet, film, rod, fiber, pipe, tubing, profile, wire coating, and extrusion blow molding. The calendering process must also be included here. Forming takes place in the nip between two rolls with the forming process generally repeated several times in increasingly finer steps. The surface will become smoother at alternating sides of the web with each passing through a nip along the whole stack of calender rolls.

Usually thermoset plastics are shaped in the pultrusion process. Continuous fiber reinforcements are first saturated with the liquid resin and then pulled through a die. The die is heated so that the resin is completely cured before the pultruded profile exits the die. Excess resin is squeezed out of the bundle at the die entrance to obtain the highest possible fiber content.

When forming thermoplastics, the final shape is not necessarily the shape of the die or the gap between the rolls. This may lead to disturbances when processing plastics into film and sheets because of the formation of edge beading. This necessitates trimming the edges so the material can be smoothly wound on to rolls. The formed extrudate can be further contoured or can be passed through a cooled calibration die. The emerging plastic products can also be oriented to a lesser or greater degree in the pull direction during drawdown. In the blown film extrusion process bidirectional orientation is achieved. A certain amount of tension must be applied at the take-off side to compensate for the die-swell occurring when the radial pressure exerted by the die land ceases. Die-swell is an expression of the plastic memory, a reaction to the

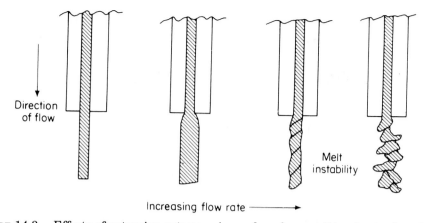

Direction
of flow

Melt
instability

Increasing flow rate ⟶

FIGURE 14.3.   Effects of extrusion rates on size and surface quality of extrudate (melt fracture). From F. Rodriguez, Principles of Polymer Systems. Hemisphere, New York, 1982.

forceful squeezing of the plastic at the die entrance. At very low flow rates (or a long die land), as shown on the left in Figure 14.3, the polymer chains have time to settle within the die dimension constraints, extruding close to the die dimensions. At the payoff side a caterpillar haul-off device or capstan take-off equipment must be properly regulated to ensure correct sizing. Control equipment is available to fine-tune the process up to 25 times/sec to ensure instantaneous, backlash-free adjustments.

These processes are ideally suited to obtain plastic items extending considerably in the length direction. Since an orientation process is achieved either concurrently or in an ensuing process, parts with higher strength in the length direction can easily be obtained. The production rates for continuous forming processes are very much dependent on cross-sectional dimensions of the formed parts. Thin fibers can be melt spun at astounding rates of 50 or more ft/sec (10 or more m/sec), whereas heavy profiles such as rods will emerge from the die at a barely noticeable speed. Extrusion rates expressed in weight per time units, however, will be much more comparable. Limitations in regard to the extrusion speed are imposed not only by the rate of cooling but also by the avoidance of melt fracture. At excessive extrusion rates the extrudate will not emerge with a smooth surface but will be rough or fractured as shown on the right-hand side in Figure 14.3. Remedies for these disturbances involve streamlining the die entrance area, lengthening the die, and unavoidably reducing the flow rate.

Frequently large numbers of cylindrical parts of different length are procured by precision cutoff operations of extruded profiles. This is, for example, the case for obtaining extrusion profiles for assorted window sizes. The greater latitude for producing parts of quite varying cross-sectional dimensions from the same die could be considered both advantageous and unfavorable. At least it substantiates the great effect meticulous process control has on extrudate dimensions.

Since extruded parts acquire their shape at a lower temperature than parts injection molded of the same plastic, their upper use temperature limit is therefore somewhat lower. Part of the orientation may reverse during heating or over prolonged time spans, resulting in some shrinkage. Excessive shrinkage may have an impact on proper progression during thermoforming of sheets and films.

Forming Against Only One Mold Surface

In many cases only one side of a shallow object or a contoured part must present a surface having a good appearance. Also frequently the part to be molded is very large or the quantities required so low that the cost for the construction of matched metal molds becomes prohibitive. In all these cases plastics parts can be produced with low cost molds having only one surface that is dimensionally exact.

These processes extend over several quite different materials and working procedures. For thermoplastics the most important processes are thermoforming, the various blow molding processes (mainly extrusion blow molding and injection stretch-blow molding), rotomolding, solution, organosol, and plastisol coating, and plastisol slush molding. Thermoset resins are mainly used in connection with fiber reinforcements. They extend over hand lay-up or spray-up moldings and the more sophisticated filament winding and robotic precision lay-up structures that are generally autoclave cured. Some of the composite processes are now also utilizing thermoplastics.

In addition to the low cost molds, in most cases these processes have the advantage of being able to adjust the thickness of the wall sections according to requirements without having to alter the mold configuration. The incorporation of stiffeners or other means of making the structure rigid are somewhat limited. In some thermoforming and blow molding processes in which a sheet or preform of certain dimensions must be used, special provisions must be met (e.g., mold plugs or prestretching) to obtain acceptable wall sections of even thickness.

Originally, all these processes had to function at low pressures or at no pressure aside from gravity. Low viscous resins were preferred for thermoset parts and low viscous organosols or plastisols for vinyl parts. To obtain coalescence of solid thermoplastic compounds in rotomolding, the lower molecular weight resins are made available in uniform particle size, free flowing powders.

As a rule, the side of the part that faces the mold surface will become the appearance side. Only with thermoforming processes, in which the starting material consists of a sheet or film with both sides equally smooth, may both sides retain a good appearance surface. Here lower quality molds with surface imperfections can still produce acceptable, somewhat more rounded contours when the reverse side becomes the exposed surface. Parts that incorporate fine details, not distinguishable from similar injection molded parts, can be obtained more effortlessly by pressure forming processes utilizing higher pressures, up to 100 psi (700 kPa), than achievable just by vacuum (Fig. 14.4).

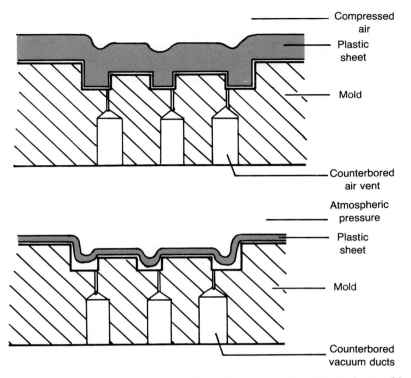

FIGURE 14.4.   Enlarged cross-sectional view of a reproduction detail obtained by pressure (top) and vacuum forming (bottom). From Gruenwald (1987).

In these processes it becomes more feasible to easily secure at least one surface of the molded part free of imperfections. For both surfaces to be free of defects matched metal molds and higher pressures have to be applied.

To obtain a strong structural part, reinforcing fibers can be rolled out to align them parallel to the mold surface. If desirable a slight pressure can be utilized after rubber bagging either by the application of vacuum or use of an elevated temperature cure at higher pressures, secured by autoclaving. The surface roughness or waviness caused by the presence of fiber reinforcements may be mitigated by the same means as described for cavity molded parts. Surfacing veil, a reinforcing mat made of very fine fibers, applied as the first layer will also improve the part's surface. Only fine mineral fillers and pigments containing gel coats will allow the production of colorful, abrasion-resistant parts having an exceptionally high gloss and smooth surface.

There are some disadvantages to these forming methods. The use of open molds will subject the top side of the laid-up part to evaporation of solvents, plasticizers, and monomers, necessitating good ventilation or recapture of volatiles. Even minor vapor concentrations of monomeric styrene and acrylates are restricted for health reasons. Furthermore many of these processes are quite labor intensive, requiring long cure times or slow heating and cooling cycles. Since the part's circumference is—by necessity—frequently smaller than the size of the mold, a costly trimming process must be incorporated.

Molecular Weight Effects on Processing and Properties

Having touched on many plastics processes and having emphasized throughout this book the importance of *molecular weight* on properties of plastics parts, these two important aspects for the production of any structural plastics part should be correlated. In Figure 14.5 the changes in properties that could be

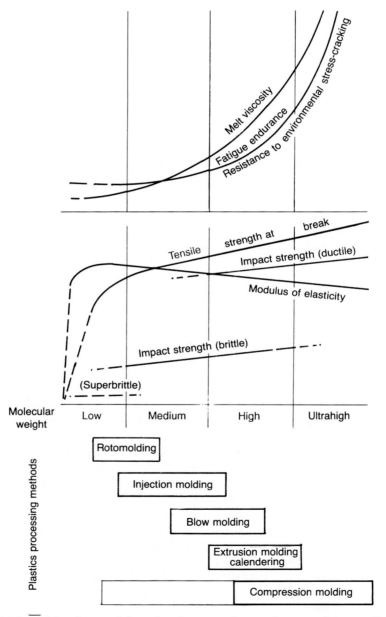

Figure 14.5. Molecular weight-related properties and processing conditions of thermoplastics.

expected to occur with increasing molecular weight are delineated. At the lower part of the graph, the molecular weight ranges are given for the more important thermoplastics processing methods. The curves relate to many commodity and engineering thermoplastics. Highly crystalline resins, high-temperature resins and their composites, and liquid crystal polymers could deviate considerably from the marked behavior.

These curves clearly illustrate that minor changes in molecular weight affect melt viscosity of all plastics and represent the main influence on any melt processing. Properties that continue to improve with increasing molecular weight are fatigue resistance, abrasion resistance, and resistance toward environmental stress-cracking. To obtain many other—mainly short time—strength properties, a much lower molecular weight polymer may suffice. In the case of modulus of elasticity, a certain reduction may take place with increasing molecular weight since alignment in the direction of stress of these excessively long polymer chains becomes more delayed. Only if these molecules are preferentially oriented (e.g., through a gel-spinning process) may an exceptionally high modulus of elasticity be reached. Impact strength properties vary in step fashion, mainly dictated by their mode of failure (ductile, brittle, or superbrittle).

Another general rule should be mentioned that relates to the establishment of *melt processing temperatures*. For semicrystalline thermoplastics about 50°F (30 K) should be added to their crystalline melting points. For amorphous thermoplastic resins about 250°F (150 K) should be added to their glass transition temperature. Since in the latter case the melt temperature will generally be lower, resin manufacturers manipulate compositions in certain classes of plastics (e.g., the polyethyleneterephthalate-type resins) to take advantage of these possible variations. These differences are also clearly expressed by the otherwise similar high-temperature crystalline polyaryletherketones and the easier to process amorphous polyarylethersulfones.

REFERENCES

M. Chanda and S.K. Roy, Plastics Technology Handbook. Marcel Dekker, New York, 1987.

J.M. Dealy and K.F. Wissbrun, Melt Rheology and Its Role in Plastics Processing. Van Nostrand Reinhold, New York, 1990.

J. Frados, ed., Plastics Engineering Handbook of the Society of the Plastics Industry, Inc., 4th ed. Van Nostrand Reinhold, New York, 1976.

G. Gruenwald, Thermoforming—A Plastics Processing Guide. Technomic Publishing, Lancaster, PA, 1987.

M.I. Kohan, Nylon Plastics. John Wiley, New York, 1973.

N.C. Lee, ed., Plastic Blow Molding Handbook. Van Nostrand Reinhold, New York, 1990.

L. Mascia, Thermoplastics: Materials Engineering, 2nd ed. Elsevier Applied Science, London, 1989.

C. Rauwendaal, Polymer Extrusion. Hanser, Munich, 1986.

R.B. Rigby, In Engineering Thermoplastics, Properties and Application. J. Margolis, ed. Marcel Dekker, New York, 1985.

D.V. Rosato and D. V. Rosato, Plastics Processing Data Handbook. Van Nostrand Reinhold, New York, 1990.

I. Rubin, Injection Molding. John Wiley, New York, 1972.

H. Simonds, The Encyclopedia of Plastics Equipment. Reinhold, New York, 1964.

J.L. Throne, Thermoforming. Hanser, Munich, 1987.

B.M. Walker and C.P. Rader, eds., Handbook of Thermoplastic Elastomers, 2nd ed. Van Nostrand Reinhold, New York, 1988.

T. Whelan and J. Goff, Molding of Thermosetting Plastics. Van Nostrand Reinhold, New York, 1990.

T. Whelan and J. Goff, Molding of Engineering Thermoplastics. Van Nostrand Reinhold, New York, 1990.

Appendix A
Bibliography

Many books and publications have been cited as references at the end of each chapter. The following listing contains either books that cover the subject of plastics and polymers comprehensively or that represent reference books, eliminating the need of citing them repeatedly at the end of various chapters.

M. Chanda and S.K. Roy, Plastics Technology Handbook. Marcel Dekker, New York, 1987.

Engineering Plastics. Volume 2, Engineered Materials Handbook. ASM International, Metals Park, 1988.

J. Frados, ed., Plastics Engineering Handbook of the Society of the Plastics Industry, Inc., 4th ed. Van Nostrand Reinhold, New York, 1976.

C.A. Harper, Handbook of Plastics and Elastomers. McGraw-Hill, New York, 1975.

C.T. Lynch, ed., Handbook of Materials Science. CRC Press, Cleveland, 1980.

L. Mandelkern, An Introduction to Macromolecules, 2nd ed. Springer-Verlag, Berlin, 1983.

H.F. Mark and N.M. Bikales, eds., Encyclopedia of Polymer Science and Technology. John Wiley, New York, 1976.

H.F. Mark, N.M. Bikales, C.G. Overberger, and G. Menges, eds., Encyclopedia of Polymer Science and Engineering, 2nd ed. John Wiley, New York, 1985.

N.G. McCrum, C.P. Buckley, and C.B. Bucknall. Principles of Polymer Engineering. Oxford Science Publications, New York, 1988.

R.M. Ogorkiewicz, ed., Thermoplastics, Properties and Design. John Wiley, New York, 1974.

W.J. Roff and J.R. Scott, Handbook of Common Polymers: Fibres, Films, Plastics and Rubbers. CRC Press, Cleveland, 1971.

S.L. Rosen, Fundamental Principles of Polymeric Materials. John Wiley, New York, 1982.

J. Schultz, Polymer Materials Science. Prentice Hall, Englewood Cliffs, NJ, 1974.

R.B. Seymour. Polymers for Engineering Applications. ASM International, Cleveland, 1987.

R.B. Seymour and C.E. Carraher, Structure-Property Relationships in Polymers. Plenum Press, New York, 1984.

A.V. Tobolsky and H.F. Mark, Polymer Science and Materials. Wiley Interscience, New York, 1971.

Appendix B
Abbreviations and Acronyms

ABA Acrylonitrile-butadiene-acrylate

ABFA 1,1-azobisformamide

ABS Acrylonitrile-butadiene styrene

ABS/PVC Acrylonitrile-butadiene styrene/polyvinyl chloride (blend)

ACS (Terpolymer of) Acrylonitrile, chlorinated polyethylene and styrene

ADC Allyl diglycol carbonate

AMMA Acrylonitrile/methyl methacrylate

AN Acrylonitrile

ANTEC Annual Technical Conference (SPE)

ASA Acrylic-styrene-acrylonitrile

ASTM American Society for Testing and Materials

ATH Aluminum trihydrate

BMC Bulk molding compound

BMI Bismaleimide

BPO Benzoyl peroxide

CA Cellulose acetate

CAB Cellulose acetate-butyrate

CaCO$_3$ Calcium carbonate

CAD Computer aided design

CAE Computer aided engineering

CAM Computer aided manufacturing

CAP Cellulose acetate propionate

CBA Chemical blowing agent

CD Compact disc

CF Cresol-formaldehyde

CFC Chlorofluorocarbon

CIM Computer integrated machining

CIP Computer integrated production

CMC Carboxymethyl cellulose

CN Cellulose nitrate (celluloid)

CNC Computer numerically controlled

CP Cellulose propionate

CPC Critical point control

CPE Chlorinated polyethylene

CPVC Chlorinated polyvinyl chloride

CS Casein

CSM Chopped strand mat

CTA Cellulose triacetate

CTFE Chlorotrifluoroethylene

DAC Diallyl chlorendate

DAF Diallyl fumarate

DAIP Diallyl isophthalate

DAM Diallyl maleate

DAOP Diallyl orthophthalate

DAP Diallyl phthalate

DBP Dibutyl phthalate

DC Direct current

DCP Dicapryl phthalate

DEHP Di(2-ethylhexyl) phthalate

Listed from *Plastics Today*, October 7, 1989, with permission of *Plastic News*, Crain Communications Inc.

DGEBA Diglycidyl ether of bisphenol A
DIDA Diisodecyl adipate
DIDP Diisodecyl phthalate
DIOA Diisooctyl adipate
DIOP Diisooctyl phthalate
DMDI Diphenylmethane diisocyanate
DMT Dimethyl ester of terephthalic acid
DNP Dinonyl phthalate
DOA Dioctyl adipate
DOP Dioctyl phthalate
DOS Dioctyl sebacate
DOZ Dioctyl azelate
DPCF Diphenyl cresyl phosphate
DPM Discrete polymer modifier
DPOF Diphenyl 2-ethylhexyl phosphate
DSC Differential scanning calorimeter
DUTL Deflection temperature under load
EA Ethylene acid (copolymer)
EAA Ethylene acrylic acid
EBA Ethylene butyl acrylate
EC Ethyl cellulose
ECTFE Ethylene-chlorotrifluoro-ethylene copolymer
EDD Engineering design database
EDM Electron discharge machine
EEA Ethylene ethyl acetate
EG Ethylene glycol
EMA Ethylene methacrylate
EMAA Ethylene methacrylic acid
EMAC Ethylene methacrylic copolymer
EMI Electromagnetic interference
EP Epoxy or epoxide
EPDM Ethylene propylene diene monomer
EPR Ethylene propylene rubber
EPS Expandable polystyrene
ESD Electrostatic discharge
ETE Engineering thermoplastic elastomer

ETFE Ethylene-tetrafluoro-ethylene (copolymer)
EVA Ethylene vinyl acetate
EVOH Ethylene vinyl alcohol
FAA Federal Aviation Administration
FCC Federal Communications Commission
FDA Food and Drug Administration
FEP Fluorinated ethylene propylene or perfluoro (ethylene-propylene) copolymer
FF Furan formaldehyde
FRP Fiberglass reinforced polyester
GP General purpose
GRP Glass reinforced polyester
HALS Hindered amines
HAO Higher alpha olefin
HCL Hydrogen chloride
HDPE High density polyethylene
HIPS High impact polystyrene
HMW High molecular weight
HMW-HDPE High molecular weight high density polyethylene
HP High performance or horse power
HPO Hydrogen peroxide
HR High resilience
HVAC Heating, ventilation and air conditioning
Hz Hertz
IM Injection molding
IPN Interpenetrating polymer network
LCD Lowest common denominator
LCP Liquid crystal polymer, polyester
L/D Length-to-diameter ratio
LDPE Low density polyethylene
LLDPE Linear low density polyethylene
LPE Linear polyethylene

MBS   Methacrylate butadiene styrene
MDA   Methylene dianiline
MDI   Methane diisocyanate (monomer)
MEK   Methyl ethyl ketone
MEKP   Methyl ethyl ketone peroxide
MF   Melamine-formaldehyde
MFR   Melt flow rate
MI   Melt index
MMA   Methyl methacrylate
MMW   Medium molecular weight
MPR   Melt-processable rubber
MUD   Master unit die
NATEC   National Technical Conference (SPE)
NFPA   National Fire Protection Association
NP   Network polymer
NPE   National Plastics Exposition (SPI)
OBPA   Oxybisphenoxarsine
OBSH   4,4-Oxybis (benzene sulfonyl-hydrazide)
OPS   Oriented polystyrene
OSA   Olefin-modified styrene-acrylonitrile
OSHA   Occupational Safety and Health Administration
PA   Polyamide (nylon)
PAA   Poly(acrylic acid)
PAI   Polyamide-imide
PAK   Polyester alkyd
PAN   Polyacrylonitrile (fibers)
PB   Polybutylene or polybutene-1
PBA   Physical blowing agent
PBAN   Polybutadiene-acrylonitrile
PBS   Polybutadiene-styrene
PBT   Polybutylene-terephthalate
PC   Polycarbonate
PC/PBT   Polycarbonate/poly-butylene terephthalate
PCTFE   Polymonochlorotrifluoro-ethylene or polychloro-trifluoroethylene
PDCP   Polydicyclopentadiene

PE   Polyethylene
PEBA   Polyether block amide
PEEK   Polyetheretherketone
PEI   Polyetherimide
PEK   Polyetherketone
PEO   Polyethylene oxide
PES   Polyethersulfone (PESV)
PET   Polyethylene terephthalate
PETG   Polyethylene terephthalate glycol (copolymer)
PETS   Plastics evaluation and trouble shooting (system)
PF   Phenol-formaldehyde
PFA   Perfluoro (alkoxy alkane)
PFF   Phenol-furfural
PGE   Planetary gear extruder
PI   Polyimide
PIB   Polyisobutylene
PIR   Polyisocyanurate
PITA   Polymer inflation thinning analysis
PMCA   Poly (methyl-n-chloroacrylate)
PMMA   Polymethyl methacrylate
PMP   Polymethyl pentane or poly (4-methylpentene-1)
PMR   Polymerizable monomer reactants
POM   Polyoxymethylene (polyacetal)
PP   Polypropylene
PPE   Polyphenylene ether
PPE/PPO   Polyphenylene ether/polyphenylene oxide
PPH   Parts per hour
PPM   Parts per million
PPO   Polyphenylene oxide
PPOX   Polypropylene oxide
PPS   Polyphenylene sulfide
PPSU   Polyphenylene sulfone
PS   Polystyrene
PSO   Polysulfone
PTFE   Polytetrafluoroethylene
PUR   Polyurethane
PVA   Polyvinyl alcohol
PVAC   Polyvinyl acetate
PVAL   Polyvinyl alcohol
PVB   Polyvinyl butyral

PVC Polyvinyl chloride
PVDC Polyvinylidene chloride
PDVF Polyvinylidene fluoride
PVF Polyvinylfluoride
PVFM Polyvinyl formal
PVK Polyvinyl carbazole
PVP Polyvinyl pyrrolidone
QA Quality assurance
QC Quality control
QDS Quality data statistics
RCF Refractory ceramic fiber
R & D Research and development
RETEC Regional Technical Conference (SPE)
RF Radio frequency
RFI Radio frequency interference
RH Relative humidity
RIM Reaction injection molding
RM Raw material
RP Reinforced plastics
RPBT Reinforced polybutylene terephthalate
RP/C Reinforced plastics/ composites
RPET Reinforced polyethylene terephthalate
RPM Revolutions per minute
RRIM Reinforced reaction injection molding
RTD Resistance temperature detector
RTM Resin transfer molding
RTV Room temperature vulcanizing
SAMPE Society for the Advancement of Material and Process Engineering
SAN Styrene-acrylonitrile
SB Styrene-butadiene
SBR Styrene-butadiene rubber
SF Structural foam
SI Silicone
SiC Silicone carbide
SMA Styrene-maleic anhydride
SMC Sheet molding compound
SMS Styrene/α-methylstyrene

SPC Statistical process control
SPE Society of Plastics Engineers, Inc.
SPI The Society of the Plastics Industry, Inc.
SPPF Solid phase pressure forming
SRIM Structural reaction injection molding
SRP Styrene-rubber plastics
SS Single stage
STAT Sheet thinning analysis (for) thermoforming
TAC Triallyl cyanurate
TCEF Trichloroethyl phosphate
TCP Tricresyl phosphate
TDI Toluene diisocyanate
TEEE Thermoplastic elastomer, ether-ether
TEO Thermoplastic elastomer, olefinic
TES Thermoplastic elastomer, styrenic
TFE Tetrafluoroethylene
TGA Thermogravimetric analysis
TiO$_2$ Titanium dioxide
TMC Thick molding compound
TOF Trioctyl phosphate
TPA Terephthalic acid
TPE Thermoplastic elastomer
TPEs Thermoplastic elastomers, styrenic
TPO Thermoplastic polyolefinic (elastomers)
TPP Triphenyl phosphate
TPU Thermoplastic polyurethane (elastomers)
TPV Thermoplastic vulcanizate
TPX Copolymer of 4-methylpentene-1
TREF Temperature rising elution fractionation
TSSC (p-) Toluenesulfonyl semicarbazide
UF Urea-formaldehyde
UHMW-PE Ultrahigh molecular weight polyethylene

UL   Underwriters' Laboratory
ULDPE   Ultralow density
      polyethylene
UP   Unsaturated polyesters
UV   Ultraviolet
VA   Vinyl acetate

VAE   Vinyl acetate-ethylene
      (copolymers)
VDC   Vinylidene chloride
VOC   Volatile organic compound
WPE   Western Plastics Exposition
WR   Woven roving

Appendix C
Stereoscopic Pictures of Crystalline Polymers

The bond angles in all organic compounds (~109°) deviate considerably from the right angle commonly used in all graphic presentations of structural objects. Therefore it becomes quite difficult to visualize the actual placement of atoms in portrayals limited to two-dimensional pictures. The next best thing to actually holding a three-dimensional atomic model is to look at stereoscopic pictures that clearly reveal the three-dimensionality of any object.

The book "Symmetry, a Stereoscopic Guide for Chemists," by I. Bernal, W.C. Hamilton, and J.S. Ricci, W.H. Freeman, San Francisco, 1972, contains many computer-generated stereoscopic illustrations showing the atomic arrangements of chemical compounds. Encouraged by reading this book, stereoscopic photographs of models representing crystalline polymers are given to the reader in this Appendix.

If no optical stereo viewing device consisting of two strong magnifying lenses is available to the reader, the two pictures can be merged into an apparently structural object by some concentration. The right eye should be positioned straight over the right-hand picture and the left eye straight above the left-hand picture. So that each eye can concentrate on its respective image, the reader must relax the eyes as if staring into the distance. If not successful at the beginning, the following maneuvers can be tried. A single corresponding atom in both images should be marked with a colored dot of about 1/4 in. diameter. This will indicate whether it will be necessary to tilt the head somewhat to make these two colored dots merge into one.

The viewer may also get very close to the paper, ignoring the fact that the picture is blurred. Then by slowly moving back to normal viewing distance, the stereo image will become sharp. As a last resort, the narrow edge of a piece of stiff cardboard, about 3 × 10 in., can be placed perpendicular to the page to separate the two pictures. When the other end of the cardboard touches the nose of the viewer, the right eye can only see the right image and the left eye only the left image. This arrangement also has the advantage of eliminating the two ghost images usually appearing to the right and the left of the stereoscopic image.

The atomic structure of diamond, positioned into the cubic unit cell, is taken from the book by Bernal et al.

1. Diamond, cubic unit cell *(1)*:

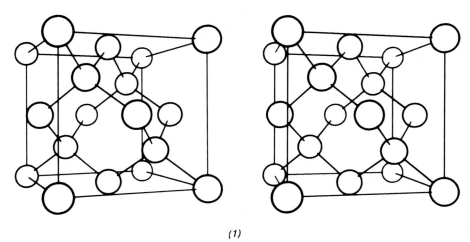

(1)

2. Comparison of *diamond* atomic structure *(2a)* with that of *cristobalite* *(quartz)* *(2b)* at comparable scale.

	Diamond	Cristobalite
Chemical formula	C	SiO_2
Average atomic weight	12	20
Specific gravity (g/cm^3)	3.51	2.32
Hardness (Mohs scale)	10	6 1/2
Fusion temperature (°C)	3827	1713

A similar spatial dispersal of atoms (formation of voids) is also found in silicone polymers if compared with polyolefins.

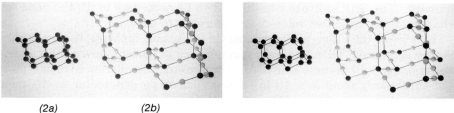

(2a)   *(2b)*

3. Segment of a *hydrocarbon chain (3)* positioned in the diamond lattice. Chain orientation of carbon atoms (black) is left to right. Carbon to carbon bonds are also highlighted in black and C—H bonds in white. Most chemical formulas containing such segments are drawn analogously throughout this book since they best describe the spatial atomic arrangements of crystalline entities of many plastics. For other plastics that exhibit a spiral arrangement, such as polypropylene, the chemical formulas are still retained in this ribbon-type arrangement due to limitations of expressing them on a flat piece of paper.

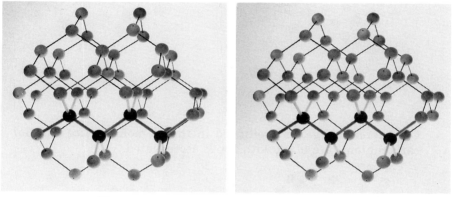

(3)

4. Representation of two prevalent atomic structures of *metals; (4a)* face-centered cubic and *(4b)* close-packed hexagonal structures.

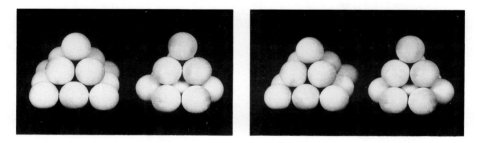

(4a)   *(4b)*

5. *Butane* molecule positioned in the diamond lattice. Solid, crystalline butane *(5a)* and only one of the many possibilities for liquid or gaseous butane *(5b)*. Butane carbon atoms are black.

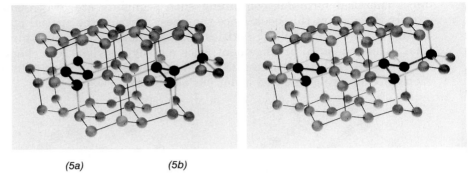

(5a)   *(5b)*

6. *Cyclohexane rings,* as positioned in the diamond lattice *(6a, 6b)*. Ring carbon atoms are black with bonds accentuated.

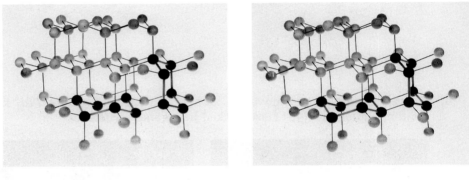

(6a)   *(6b)*

7. *Lonsdaleite* atomic lattice shown with the two different cyclohexane rings (chair and boat form) *(7a, 7b)* outlined in black with ring bonds accentuated.

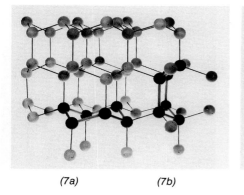
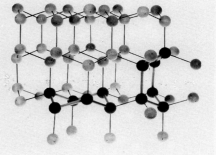

(7a)   (7b)

8. Section of *paraffin* crystal positioned in the lonsdaleite (diamond) crystal lattice in which all spheres represent carbon atoms *(8)*. Only the carbon atoms of the superimposed paraffin molecules are black and their C—C bonds are highlighted. Hydrogen atoms are white.

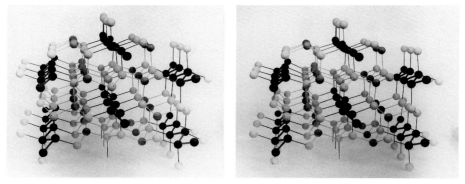

(8)

9. Regular chain folding in *polyethylene* single crystallite *(9)*.

(9)

10. *Isotactic polypropylene* combined with the diamond lattice (sagittal direction) *(10)*. Carbon atoms are black. Chain links are accentuated. Hydrogen atoms are white.

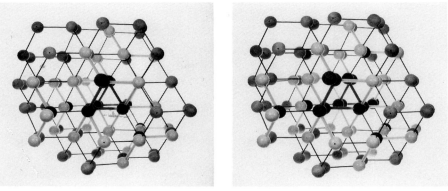

(10)

11. *Isotactic polypropylene,* perpendicular view (vertical direction) *(11).*

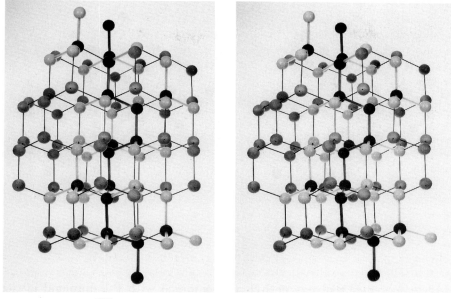

(11)

12. *Syndiotactic polypropylene* combined with the diamond lattice (sagittal direction) *(12).* Carbon atoms are black. Chain links are accentuated. Hydrogen atoms are white.

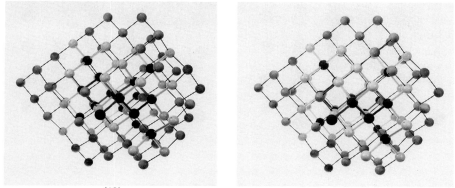

(12)

13. *Syndiotactic polypropylene* perpendicular view (vertical direction) *(13)*.

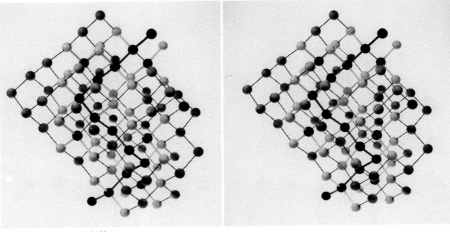

(13)

14. *Acetal* polymer (polyoxymethylene) combined with the diamond lattice (sagittal direction) *(14)*. Carbon atoms are black, oxygen atoms are gray, and hydrogen atoms are white. Chain links are accentuated.

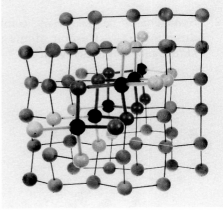

(14)

15. *Acetal* polymer (polyoxymethylene), perpendicular view (vertical direction) *(15)*.

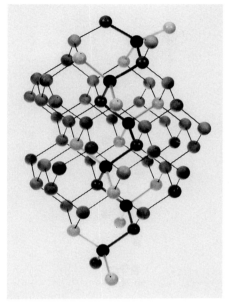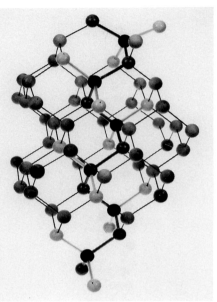

(15)

16. *Cellulose* structure *(16)*. Carbon atoms are white, oxygen atoms are gray. Hydrogen atoms not shown.

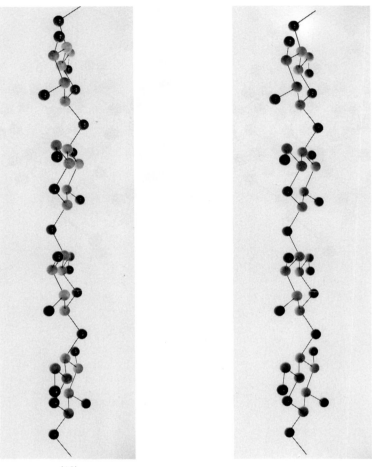

(16)

17. *Collagen* model *(17)* symbolizing helix formation of three polymer chains.
     Flags indicate side rings of two out of the three amino acids.

(17)

Index

T Data is displayed on Tables; **bold** indicates main subject treatments on one or more pages. **Acronyms** are listed separately on pages 320–326.